WAYS OF NATURE

WAYS OF NATURE

HOW THE TRAILS OF LIFE EXPOSE THE UNIVERSE

PANY DECOSSARD

Tampa, Florida

Ways of Nature: How the Trails of Life Expose the Universe

Published by Gatekeeper Press
7853 Gunn Hwy., Suite 209
Tampa, FL 33626
www.GatekeeperPress.com

Notice of Risk and Disclaimer

Exodus 3:14 quotation is from the Holy Bible, Catholic Public Domain Version (CPDV).

Isaiah 44:6 quotation is from the King James Version (KJV) and extracted from Gutenberg.org.

Library of Congress Control Number: 2022936713

ISBN (hardcover): 9781662927454
ISBN (paperback): 9781662927461
eISBN: 9781662927478

To my wife, Michaelle,
and our daughters, Régine and Jasmine.

To my mother, Aménise, née Foucault.
And to the memory of my dad, Léger.

We see but a part, and fancy that we have grasped the whole.

—Empedocles

Table of Contents

Part I: On the Origin and Evolution of Eukaryotes

Part II: On the Origin of Cells and Viruses

Part III: On the Commonality of the Inert and the Living

Preface

There was a time in the history of humanity when diverse aspects of reality were explained by the power of different gods or spirits. Then, progressively crept into the conscience of some of our forebears the idea of an omnipotent God, the "one God," to whom was attributed the exclusive creative power that accounts for the whole universe. With some minor variations, this is the God implied or named in the theologies of thinkers such as Plato, Aristotle, or Thomas Aquinas, and the God of the Abrahamic religions. The idea of God was a philosophical or religious "theory of everything" (TOE). The proposition of a primary or "uncaused cause" at the origin of everything may have been the greatest philosophical insight of all time. Eventually, others started to realize that with the lessons learned from empirical observations, such as the movements of "heavenly" bodies, they could make certain predictions, such as the timing of eclipses or derive some laws of nature. In other words, natural phenomena could be explained or accounted for by natural causes. Slowly but surely, science and its allied philosophy, methodological naturalism, entered the scene.

The work of science is usually done away from the limelight, and often the name of a scientist, even after a lifelong career, does not go beyond the walls of his laboratory, or at most a very small circle of people with the same interests. But from time to time, major breakthroughs or revolutions carry through space and time the names of those involved. Copernicus, Newton, Darwin, Einstein, Watson and Crick, Doudna and Charpentier* are just a few. It remains no less true that through all these efforts, the timeless yearning for a single principle that could explain the universe has remained the holy grail, or rather, the holy of holies of the scientific enterprise. Einstein is said to

* The recent discoveries and advances in gene-editing technologies involving CRISPR-Cas9 are also counted among these defining moments in science.

have spent the last thirty or so years of his life unsuccessfully searching for such a formula, or at least one that would reconcile quantum mechanics with his general theory of relativity. Soldiers of science have been waging an offensive war to conquer inches by inches the vast land gifted to God by the prophets and the philosophers. Their aim is to force the idea of God to retreat, shrink the accepted sphere of the divine to the rim, and drive it, so to speak, to whittle away. They would then either discover God on their own terms or replace Him with their own TOE. All scientific undertaking is seen through this filter as an endeavor to win even a little more support for naturalism.

This project, however, which started as a quite aimless scientific stroll, eventually took on a life of its own. An intellectual foray into the philosophy of reproductive biology has brought me to write this book on life and its evolution, in the grand scheme of the nature and origin of our universe, subjects on which I had only what I considered a rudimentary grasp. Suffice it to say that, as a physician, I have dipped my toes in not-so-familiar waters to grapple with the many interrogations and challenges brought on by this journey. But each time, keen insights accrued from a career in the life sciences—and perhaps also from a childhood in the countryside, with a deep involvement in the farm life of my parents—have, I think, pulled me out of trouble. As implied by Gregor Mendel's character in the *NOVA* episode "The Garden of Inheritance," the processes of a farm can effectively teach a young mind some important rudiments of genetics.

This book is a wide embrace of biology, physics, and perhaps a bit more; one that shines a brand-new light on the controversial topic of the origin and evolution of life, and at the same time takes a novel and original look at cosmogony. It will attempt to answer some fundamental questions on the origin of eukaryotes, their evolution, and the destiny of humanity. I wrote the book without a specific reader in mind, but all the while trying not to dilute the best evidence supporting my theories. At times, the discussion may seem too specialized for some, but I usually return to a more narrative approach. The reader may feel free to skip the specialized parts without losing too much, as the themes are most often recurrent in different forms throughout the book. One may see that, on occasion, I tend to repeat myself without much care for elegance. Here I will shamelessly attempt to justify this peculiarity of mine on my flunking the course on the French poet Boileau,* and on guidance first provided by Ludwig Boltzmann but later revived by Einstein[1] that elegance should be left to tailors and shoemakers.

In any theory, as in the conclusion of any scientific experiment, there is always a certain degree of uncertainty. This book is no exception. First of all, I have done no empirical experiments. Second, flaws, misinterpretations, and incomplete explanations will be inevitable, due in part to the ambitious and eclectic nature of the project. But they will also owe a great deal to our bold conviction that our lack of credentials should not deter us from trying. One can only take solace in the fact that no single person can claim expertise in the totality of the subjects addressed in this book, for they are multidisciplinary and wide-ranging.

Having seen this project through, I feel as though I had been invited to try a maze and succeeded to work my way to the exit, just as Theseus managed to reach the Minotaur in the Cretan labyrinth. However, I humbly admit the infinitesimal value of my knowledge of the immensely vast cornfield. In that regard, I take a position in line with Darwin's when, to address the warning of the Scottish surgeon Hugh Falconer that posterity might erect a new superstructure upon the foundation of his theory of natural selection, he replied,

---

* "*Tout ce qu'on dit de trop est fade et rebutant;*
L'esprit rassasié le rejette à l'instant.
Qui ne sait se borner ne sut jamais écrire." —Nicolas Boileau, in *L'Art poétique.* (Summary translation: Whoever cannot limit himself never knew how to write).

"I look at it as absolutely certain that very much in the *Origin* will be proved rubbish; but I expect and hope that the framework will stand."[2] While in my case, I feel, obviously without coyness but neither with hubris, even more optimistic than Darwin, I recognize that the claims I make here may not all carry an equal measure of plausibility. Yet, in the soundness of all those I have made, I believe.

This book would not have seen the light of day if not for the support of my dear wife, Michaelle, the pillar of our home. Thank you, Mich, for your unwavering support. My thanks go as well to my mother, Aménise, who set me on the first steps of my much longer journey. This book is dedicated to you both.

Warwick, NY
February 16, 2022.

PART I

On the Origin and Evolution of Eukaryotes

CHAPTER 1

Birth of the Domain Eukaryota: Theory of Bilateral Descent

It is owing to wonder that men both now begin, and at first began, to philosophize.

—Aristotle

Determinism of Biological Reproduction

The sudden realization of this fly, trapped as it was, right in the ointment of nature, struck me like a lightning bolt. Or rather, it occurred to me that this was something in dire need of further scrutiny and investigation. The ubiquitous phenomenon of reproduction, and most especially sexual reproduction, I sensed, must conceal an even more extraordinary story to tell us about life. Then, by pure serendipity, out came some of the insights recorded in the following pages.

Indeed, careful observations of the behaviors and processes of life forms can lead to meaningful eureka moments. A bacterium finds its way to our subcutaneous tissue to multiply, form an abscess, or uses genetic transfer mechanisms to share a fragment of its DNA with a different bacterium. A virus of the common cold commandeers its replication by hijacking the reproductive machinery of mucosal cells of the human upper airway. Plants evolved varied creative methods to disperse their seeds to new grounds. One watches with awe the dandelion or the milkweed seeds harnessing the power of the wind with their flying apparatus, taking off to disseminate, find new fertile soil, and reproduce. The young teenager trekking at serious personal risk to far-flung war theaters to meet a potential mate or spouse is nothing short of bewildering. It is similarly with consternation that one watches the male lion who, making way for his own genes, kills the unweaned cubs of the pride he just took over.

From a different perspective, the mother bacterium ceases to exist after the binary fission proceedings have given birth to two new daughter cells. Annual plants such as grains die after one reproductive cycle. Equally compelling is the phenomenon of menopause in women. What is the economy of the progressive health decline heralded by this biological stage? The lack of estrogen brings about a wide variety of physical and functional changes in the woman's body that not only decreases its attractiveness to men but also marks a turning point in its aging process. The postmenopausal progressive genital atrophy and increased incidence of potentially life-terminating conditions such as cancer and cardiovascular diseases seem revealing. One is also puzzled by the semelparous reproductive pattern of various species, like the Pacific salmon (*Oncorhynchus tshawytscha*), the drone bee (*Apis mellifera*), the deer tick (*Ixodes scapularis*), and many other animals in which death is programmed during the course of reproduction.

One can infer that all those processes are inherent to the reproductive determinism of the genome carried by the different organisms mentioned. As if the marching orders of the individual organism were, "Reproduce and die!" Sheer doggedness, while apparently a trait of the biological lineages, is not one for the individual organism who merely survives long enough to reproduce and raise its offspring if necessary. Living systems are animated by a natural or instinctive drive to transmit their genome to a new generation. On the other hand, it goes without saying that proliferative conditions represent, at the cellular level, a corollary of this process running amok. Based on this assessment, and drawing from the 1994 NASA definition of life,[1] one could define natural living forms as chemical systems capable of directing the reproduction and transmission of their genome. It is worth noting, however, that viruses, which are usually not considered living, fit that definition. On the other hand, we can all agree that the genetic hybrid animals such as mules, hinnies, and zebra/horse crosses belong to the living world, despite their usual infertility. Yet, their infertility arises not from a lack of reproductive instinct, but mainly because they are unable to form functional gametes secondary to the chromosome discrepancies or imparity of their parents. In scientific literature, this interpretation is referred to as the "Dobzhansky–Muller (D-M) incompatibility" model.[2] More precisely, according to this model, hybrid sterility or inviability results from the functional incompatibility of a certain portion of the genetic material (at least two interacting genes that are unable to cooperate).[2-4]

Therefore, reproductive capacity—defined here as the ability to transmit one's genome to a new generation—does not seem to be a *sine qua non* criterion for the otherwise normally functioning cell or multicellular organism to be considered living. Enough to say that life still needs a valid and comprehensive definition, one that includes every living system and clearly excludes inert materials. Bacteria constitute the prototypical life form, being self-sufficient and capable of reproducing individually, provided the existence of a suitable medium and the necessary organic nutrients. In this context, infectious diseases of a given species practically represent collateral damage suffered because of the reproductive processes of less complex life forms. Viruses need the intracellular reproductive machinery of this or that cell type, while different bacteria or parasites prefer different milieux or biosystems of an organism. The more complex the organism, the more the potential for it to be taxed for the reproductive needs of simpler life forms. At the same time, one should not forget to mention our symbiotic relationship with our microbiota and our larger ecosystem, a relationship that, when well-balanced, is neither adversarial nor detrimental to the host.

As it relates to living systems, the term "reproduction" is used here only metaphorically, or by extension. While a bacterium or some viruses may produce offspring that are exact copies of themselves, in unisexual eukaryotes, one parent only contributes half of the genome of its offspring.

Our Ancestors, Hiding in Plain Sight

The Bilateral Descent Theory of Eukaryogenesis

This epiphany led us to ponder the way in which this imperative of reproduction of the nucleotides and of the organisms that serve as their vehicles corresponds to the genesis and evolution of eukaryotic lineages. We will start from the conviction that our present phenotypes have resulted from an evolutive process since the apparition of eukaryotic life, a process that includes both the information comprised in our genomes and data from our specific ecosystems. We then propose that, to discover our most distant eukaryotic ancestors, we ought to strip ourselves of some of the features of our evolutionary shell, be it the horns or the tusks, the wings or the arms, the pigmented skin, the blue irises, etc. If we bring ourselves down to our bare, naked first cell, the zygote, we will have a better glimpse of our origins. The zygote, born of the fusion of the ovum with the sperm cell, is the first phenotypic expression of a eukaryotic organism. We posit that the gamete cells are positioned among our earliest eukaryotic ancestors. The characteristic fertilization mechanism of the animal species, for instance, is represented by a flagellated male gamete cell injecting its genetic material into an egg, like a virus, namely a bacteriophage, infecting a bacterium. We propose that eukaryotes emerged from the fusion of bacteriophage viruses or phages with bacteria. This is the essence of the viro-bacterial or bilateral descent theory of eukaryogenesis.

RNA or DNA Phage?

Our most remote "male" ancestor was a bacteriophage. What type of bacteriophage was it? Was it an RNA or a DNA phage?

Mammalian sperm cells are flagellated dsDNA cells. They show some morphological similarities with the Caudovirales, all of which are tailed dsDNA phages and are believed to share ancestors with common features. However, we propose that the entities that played the role of the pioneer "sperm cells" were (at least originally) bacteriophage retroviruses or "retrophages." RNA is thought to be chronologically the first molecule of heredity.[5] It has been a surprise to find that a large fraction of most eukaryotic genomes, about half in humans,[6] derive from retrovirus-like elements called retrotransposons. It has been proposed, and we believe also, that the modern predominance of dsDNA phages and other DNA viruses result from an evolution away from a world where RNA was the sole genetic material of living systems.[7,8] We are submitting that the pioneer sperm cells were temperate retrophages whose RNA information underwent reverse transcription within the bacterial hosts before the distinctive genome of the individual protagonists on both sides of the bilateral descent system combined into one for the new eukaryotic cells. This pioneer retroviral heritage is widely dispersed across eukaryotic genomes, some of its signatures being the endogenous retroviruses (ERVs) and the retrotransposons. It did not invade our germline; it is an integral part of it because it constituted half of the diploid genome of our first cell as eukaryotes.

In this context, it is also noteworthy that the capping and polyadenylation patterns of retroviral genomic RNAs and of eukaryotic mRNAs are identical.[9] Along the same lines, the phenomenon of viral superinfection exclusion (SIE) was an early form of immunity and the precursor of polyspermy prevention in eukaryotic fertilization. Superinfection exclusion is a phenomenon in which an ongoing viral infection renders its target, a cell or an organism, refractory to a *de novo* infection by another identical or a closely related virus. It is a virus-controlled function mediated in each case by a specific viral protein.[10] For instance, SIE by the Citrus tristeza virus (CTV), an RNA virus, is mediated by the viral p33 protein.[10] Some elements of comparison between retroviruses, mammalian sperm cells, and the tailed-bacteriophages or Caudovirales are presented in Table 1.

Table 1. Elements of Comparison between Retroviruses, Caudovirales, and Mammalian Sperm Cells

Guest	Method of Entry into Host	Host Immunity	Genome	Appendage
Retrovirus	Fusion of membranes[9]	Superinfection exclusion (SIE)*	RNA	N/A
Caudovirales	Tail-mediated penetration of bacterial wall[12]	Superinfection exclusion**	dsDNA	Tail
Mammalian Sperm Cell	Fusion of membranes[14]	Polyspermy prevention***	dsDNA	Flagellum

*ex: mediated in HIV by Nef protein.[11]

**ex: in temperate *Streptococcus thermophilus* phage TP-J34, mediated by Lipoprotein Ltp.[13]

***mediated by ovastacin in mammals.[15]

What About the Egg?

Whereas the sperm is essentially genomic information sheltered within a motile nucleus, the human ovum is a nonmotile, comparatively large cell, harboring a uniquely complex and abundant cytoplasm.[16] Which bacterium, then, is the precursor of the egg? I was pondering this question when I came to know about the discovery in 1999, off the coast of Namibia, of the world's largest bacterium by Heide Schulz of the Max Planck Institute for Marine Microbiology and her team. This gammaproteobacterium, named *Thiomargarita namibiensis*, which has been described elsewhere,[17] or probably an antecedent organism, appears to have all the characteristics of the egg ancestor of the animal kingdom. Henceforth, we will often refer to this bacterium simply as *Tn*.

Other sulfur- and nitrate-rich bacteria, such as those of the genera *Thioploca* and *Beggiatoa,* also belonging to the class of Gammaproteobacteria have been found and described.[18] Like *T. namibiensis*, they are large bacteria packed with metabolic substrates that allow them to potentially act as incubators in addition to their ability to transmit genes. *Thioploca* and *Beggiatoa* bacteria appear, however, to be prokaryotic precursors of algae and the ova of plants by the same mechanism. The gliding motility[17] they display toward their substrates (sulfur and nitrate) can be interpreted as a forerunner of heliotropism and plant root formation. Besides, evidence of carbon dioxide fixation and the presence of photosynthetic enzymes of the Calvin-Benson cycle have been reported in those bacteria.[19-21] Some similarities between animal oocytes and *T. namibiensis* are described in Tables 2 and 3. We, therefore, submit that *Tn* or a *Tn*-like bacterium is the ancestor of the egg of the animal kingdom. Likewise, we cannot

Table 2: Descriptive Comparison between *T. namibiensis* Bacteria and some Eukaryotic Eggs

Cells	Size	Shape	Visibility to the naked eye	Motility	Larger determinant of biovolume	References
***Tn/Tn*-like**	100-300 µm	Spherical	Yes	Nonmotile	Large vacuole (98% of biovolume)	17
Mammalian oocyte	60-100 µm in general. Human oocyte: 100 µm	Spherical	Yes	Nonmotile	Large cytoplasm	22-27
Eggs of oviparous animals	Large, and vary in size	Ovoid, or roughly spherical	Yes	Nonmotile	Large nutrient reserve/yolk. (Over 95% of biovolume in some species)	23, 24

Table 3: Metabolic Comparison between *T. namibiensis* Bacteria and some Eukaryotic Eggs

Cells	Environmental oxygen (A)	Use of nitrogen and sulfur-containing substances (B)	Excretion of nitrogen-derived compounds (C)	References
***Tn*/*Tn*-like**	Low (a)	-Accumulate nitrate and sulfur -Use nitrate as electron acceptor to oxidize sulfide for energy (sulfur/nitrogen cycles coupling) -Facultative aerobic. May use oxygen as an electron acceptor in the presence or absence of nitrate	Ammonium production through dissimilatory reduction of intracellular nitrate (b)	**A/B** 17, 28 **C** 18, 28-30
Mammalian oocyte*	Relatively low oxygen concentrations in mammalian oviducts (c)	Mammalian oocytes and early embryos use a multiplicity of energy substrates (d). However, sulfur-containing substances such as taurine, hypotaurine, and sulfated macromolecules have been found to concentrate in the female genital tract fluids of several mammals (e). There is evidence for endogenous production and utilization of nitric oxide and hydrogen sulfide by oocytes and early embryos of some mammals. Same gas also produced in female reproductive tissues (f).	Ammonium generated in culture media of mouse and preimplantation human embryos (g)	**A** 31-33 **B**(d) 34-37 **B**(e) 38-42 **B** (f) 43-53 **C** 54-56
Oviparous animal eggs	The egg and embryo develop inside a shell in an oxygen-deprived environment	- Hen's eggs accumulate nitrogen and sulfur-rich amino acids such as cystine, methionine, and taurine - Sulfur-containing amino acids, especially methionine, are the first limiting nutrients in typical laying hen diets - Dietary taurine supplementation promotes egg production in laying hens and Japanese quail	Compounds such as ammonia, urea, and uric acid are excreted by oviparous animal eggs and embryos	**B** 57-64 **C** 65-68

*The complete metabolic pathways of mammalian oocytes and early embryos have not yet been fully elucidated.

a- *Tn* found in low O_2 concentration environment, 0-3 μM (Micromolar).

b- *Tn* and the filamentous sulfur bacteria of the genera *Beggiatoa* and *Thioploca* are closely related and seem to have similar physiology. They produce ammonium through dissimilatory reduction of intracellular nitrate.

c- Existing data indicate that early embryos develop in vivo under low oxygen tension, particularly during the peri-implantation period. Relatively low oxygen concentration in mammalian oviducts; 1.5–9% in rhesus monkeys, hamsters, and rabbits, vs. 20% in the atmosphere. An O_2 concentration of 5% in the culture media is reported to improve embryo development for multiple species.

d- Mammalian oocytes and early embryos have been reported to use a multiplicity of energetic substrates that include pyruvate, glucose, nonessential amino acids, etc.

e- Taurine was found to support the preimplantation development in vitro of human and mouse embryos.

f- Nitric oxide (NO) and hydrogen sulfide (H2S) have been shown to play an essential role in early embryonic development and the establishment of pregnancy in multiple species.

g- Currently purported mechanisms include the metabolic degradation of amino acids in addition to their spontaneous deamination.

help but notice that the similarities and the proposed parentage between *Thiomargarita namibiensis* bacteria and mammalian oocytes indicate the possibility of improving the success rate of IVF culture media for mammalian embryos by modeling them on the metabolic needs of these bacteria and the characteristics of their natural environment. Additionally, based on our findings, egg production in laying hens can also be increased when fed a diet inspired by such an approach.

Thus, specific gammaproteobacteria (*Tn* or a *Tn*-like bacterium in the case of the animal kingdom, those of the genera *Thioploca* and *Beggiatoa* in plants), through lysogenic infection by respective bacteriophages, are hypothesized to have formed the first eukaryotic cells. The infected bacteria will be transformed into eukaryotic cells and effectively become fertilized eggs. Eukaryotes are either unicellular or multicellular. Multicellular eukaryotes are also called metazoans. It is assumed that some zygotes resulting from the fusion of specific bacteriophages and their corresponding bacteria have remained unicellular and given rise to unicellular eukaryotes. Unicellular eukaryotes, by and large, reproduce asexually by budding, binary fission, or mitosis. Other zygotes resulting from virus-bacterium pairs will form the multicellular eukaryotes, namely plants, animals, and most fungi. The pioneer metazoans will add gamete fertilization as a new reproductive strategy to their repertoire by the establishment of meiosis, a novel modality of cell division. Meiosis will lead to the first apparition of proper male and female gamete cells, and sex eventually will become one method to rendezvous those two protagonists of fertilization.

In this way has emerged the world of eukaryotes!

In this way, we were born!

Moreover, we also call this viro-bacterial theory of eukaryogenesis the theory of bilateral descent, in accordance with the tribal structure of some indigenous peoples of Africa and elsewhere, of which the OvaHimba tribe of Namibia and Angola is arguably one of the best known. The bilateral descent system is one in which each tribe member is part of two equally important clans: the paternal (patriclan) and the maternal (matriclan).[69] Under the tribal bilateral descent system, wealth and property (mainly cattle) are inherited only from the matrician clan. For instance, a man inherits his maternal uncle's livestock but not his father's. This inheritance structure mirrors perfectly well the case of the zygote, whose cellular resources are derived exclusively from the egg or originally from the bacterium: the contribution of the retrovirus and subsequently of the sperm consisting only of half of the genetic material. What is now called the specific endogenous retroviral heritage of each species stems from this original retroviral contribution to their first eukaryotic cell.

Creation of the Eukaryotic Cell

Until now, the appearance of eukaryotic cells on a global scale has remained one of the greatest enigmas of biology. Eukaryotes constitute one of the three domains of life as defined by Carl Woese.[70, 71] Bacteria and Archaea, which together make up the group of prokaryotes, are the other two domains. Using cellular differences in ribosomal RNA sequences, Woese built a phylogenetic tree of life representing the three domains described above, but that obviously excludes viruses (viruses are not cells, and they do not have ribosomes). The model depicts an evolutionary tree where the three branches sprang up through DNA mutations, according to the neo-Darwinian framework, from a common ancestor.[71] Stemming from a hypothetical last universal common ancestor or LUCA, one branch would lead to Bacteria, while another would split to give Archaea and Eukaryota. (Fig. 1).

The eukaryotic cells are usually defined in opposition to those of prokaryotes, due to several distinctive features.[72] They are more complex and of a much larger size than the latter. Furthermore, the eukaryotic

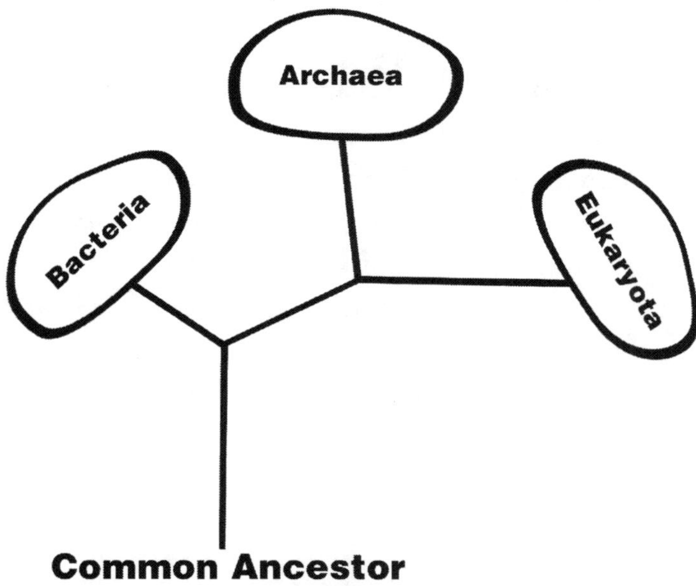

Figure 1. The three domains of life, according to Carl Woese.

cells dispose of a few exclusive features, the most prominent of which is the presence of intracytoplasmic membrane-bound organelles such as mitochondria, chloroplasts, and a nucleus. It is generally assumed that eukaryotic cells arose from prokaryotes. We instead submit and specify that eukaryogenesis resulted from the fundamental transformation of some gammaproteobacteria after a retrovirus bacteriophage infection through genome hybridization and horizontal gene transfer. Such an event obviously punctuated, if you will, a previous biological equilibrium of the involved bacteria. The genetic contribution of the retrophage transforms its gammaproteobacterial host into a eukaryotic cell, a new domain of life.

Eukaryogenesis involves the lysogenic infection of a gammaproteobacterium by a retrophage. The gammaproteobacteria are among the largest known bacteria.[17, 73] It ensues that the eukaryotic cells inherit from them their relatively large diameter. Besides, the compartmentalization of DNA replication away from the cytoplasm in a nucleus is a key feature that distinguishes eukaryotic cells from prokaryotes. Vorrapon Chaikeeratisak and collaborators have reported the assembly of a nucleus-like structure during the replication of a bacteriophage in *Pseudomonas aeruginosa*.[74] In that same vein, more recently Senén D. Mendoza and colleagues demonstrated that ΦKZ bacteriophages infecting *Pseudomonas* were able to protect their DNA from the cleavage action of CRISPR-Cas and restriction nucleases by shielding their genetic material during replication in a similar nucleus-like compartment.[75]

We believe that when the bacteriophage fused with the gammaproteobacterium to form the eukaryotic cell, the virus provided the nucleus to the new structure. The nucleus fulfills for the new cell, which contains the combined viral and bacterial genomes, the same function of genomic protection the capsid afforded to the viral genetic material when it was single, and much more. Likewise, James A. Kraemer and collaborators have described a novel family of divergent tubulins, named "PhuZ," encoded in the genomes of phages and which position their DNA in the center of the host cell.[76, 77] Tubulins play a critical role in DNA segregation during cell division in eukaryotes. While bacteria contain some microtubular structures, they generally do not have *bona fide* tubulins, a feature of the eukaryotic cell.[78] The nucleus and elements of the eukaryotic cytoskeleton represent part of the contributions of the phages themselves. Moreover, in most species, including humans, the sperm-derived pronucleus provides the centrosome during fertilization, a situation that is very evocative of the contribution of tubulins by the phage during eukaryogenesis.[14]

Another question our theory must address is the origin of the chloroplasts and mitochondria of eukaryotic cells. Chloroplasts are chlorophyll-containing organelles of algae and plants, the sites of photosynthesis or the conversion of solar radiation into chemical energy in the form of adenosine triphosphate (ATP). Their origin is obviously traced back to cyanobacteria. Mitochondria, on the other hand, are the respiratory organelles of virtually all eukaryotes.* They are the centers of oxidative phosphorylation where the coupling of cellular oxidations with the assembly of ATP takes place. The compartmentalization of respiration in mitochondria occurs within the folded expansions of their inner membrane, called cristae.[72] Those mitochondrial cristae were found to be homologous to the intracytoplasmic membranes (ICMs) used by diverse alphaproteobacteria to harness energy. This assessment is supported by the finding of sequence and structural similarities between mitochondrial cristae organizing systems (MICOS) and the related system of alphaproteobacteria (alpha MICOS). The argument that mitochondrial cristae originated from the SAR11 clade of *Alphaproteobacteria*, the one whose bacteriophages are called pelagiphages, is therefore quite strong.[80-83]

The current proposition to explain the presence of those organelles inside the eukaryotic cell is the symbiogenesis or endosymbiotic theory. This theory has a long history; but in more recent times, it was developed and advocated by the American biologist Lynn Margulis.[84] It holds that those two eukaryotic organelles originated from former free-living prokaryotes that were engulfed by the ancestors of eukaryotic cells. For our theory to account for this view, every eukaryogenic bacterium, either before or after pairing with a virus, would have to acquire its own endosymbiont(s). However, to explain eukaryogenesis under a unified mechanism in conformity with the principle of parsimony (Occam's razor), we submit that the newly formed eukaryotic cells (formerly eukaryogenic bacteria) obtained the copy of the genes of the photosynthetic and respiratory organelles by relevant bacteriophage-mediated horizontal genetic transfer (HGT) from cyanobacteria and alphaproteobacteria, respectively. The mechanism of bacteriophage-mediated genetic

* No mitochondria have been found in the eukaryote *Monocercomonoides sp.*[79]

exchange between bacteria is referred to as transduction and was discovered by Norton Zinder and Joshua Lederberg in 1952.[85]

The eukaryogenic bacteriophages may have transferred the genes of those organelles into their gammaproteobacterial hosts as plasmids. For instance, phage-mediated transduction of both chromosomal and plasmid DNA has been reported in diverse natural environments.[86] In fact, the draft genome sequence of a sample of the sulfur bacterium, *Beggiatoa* ("*Candidatus Maribeggiatoa*"), has shown evidence of extensive genetic exchange with cyanobacteria, putatively through mobile elements, as reported.[87]

This requirement of eukaryogenesis infers that before the definitive fusion of eukaryogenic bacteria with their bacteriophages, the latter would have acquired the genes necessary for the organelles in question by infecting other species of bacteria, an apparent challenge for our theory, given the species-specific nature of bacteriophage infections.[88]

However, it has been found that some phages can devise creative solutions to develop polyvalence, i.e. the ability to infect more than one species.* To that end, the phage *Mu*, for instance, has the means to significantly modify its receptor recognition system.[88] Similarly, the remarkably numerous and diverse marine viruses are found to be much involved in the phenomenon of lateral genetic transfer. For instance, genomic sequencing of these viruses has revealed a large number of genes that were otherwise involved in cellular metabolism. Case in point, photosynthesis genes were found to be common in

---

* We have not determined whether the proposed infection of cyanobacteria and alphaproteobacteria by eukaryogenic bacteriophages was done during their lifespan as retroviruses, or rather as DNA bacteriophages before the final eukaryote-producing fusion with gammaproteobacteria. Each of those two possibilities has its own set of implications for the host bacteria and their viral "progeny" (for example, lytic or non-lytic replication, enveloped or non-enveloped virions).

cyanophages (phages of cyanobacteria).[89] Again, in that regard, polyvalence is not an unusual feature of cyanophages.[88]

The novel domain of life, called Eukaryota, while presenting some defining characteristics (cells with membrane-bound organelles and of relatively larger size, etc.), was not homogeneous, however. Several eukaryotic cells arose from different virus-bacterium pairs. As far as the identification of the bacteria involved in the genesis of the eukaryotic cells, we can only state at this juncture that *T. namibiensis* gave rise to the animal kingdom, while *Beggiatoa* and *Thioploca* bacteria were involved in the emergence of algae and plants. As for the bacterial ancestors of fungi, we cannot at this time make a statement one way or the other. We did not consider this question with any intensity.

CHAPTER

2

Emergence of Animals and Creation of Meiosis in Eukaryotes

Fate of the First Eukaryotic Cells

Eukaryotes can be classified in many ways. One of them is their cellularity or the number of cells in an organism. Cellularity is categorized in a binary fashion as unicellular or multicellular. For example, most protists, including protozoa and some algae, are unicellular or single-cell eukaryotes,[1] while plants and animals make up the major representatives of multicellular organisms.

Several lines of evidence may be able to support the contention that the most remote multicellular ancestors of the animal kingdom expressed the sponge phenotype. Some South Australia Ediacaran specimens of the Neoproterozoic era have been identified as sponges.[2,3] In that regard, the difficulty in interpreting some of the microscopic structures preserved in the Doushantuo Formation of Southern China as either eggs and embryos of early animals or *T. namibiensis* bacteria is understandable.[4, 5] It is because both interpretations are correct. We believe that those specimens are fossils of fertilized *Tn* bacteria in the process of their transformation into embryos of the sponge phenotype. The research conclusion of a team of scientists from the Massachusetts Institute of Technology (MIT) seems to provide further support for the assessment that animals first displayed a sponge phenotype.[6] An unusual molecule, the sterane 24-isopropylcholestane, which has been suggested as a molecular biomarker for sponges, was found in 640-million-year-old rocks. Currently, certain sea sponges synthesize 24-isopropylcholesterol (24-ipc), the likely precursor of this molecule.

Thus, in all likelihood, animals emerged phenotypically as sponges (phenotype of the phylum Porifera) at the bottom of the hydrosphere. While our modern phenotypes are a far cry from their filter feeder days after close to a billion years of evolution, sponges still exist in the hydrosphere, at the bottom of the oceans, and in fresh water as well,[7] as living fossils of a long-lost world. Sponges reproduce both sexually and asexually.[7-9] Their asexual reproductive strategies consist of a variety of methods, such as external budding, gemmulation, and fragmentation. External budding occurs when an outgrowth or bud regenerates into a separate organism after breaking off from the parental sponge. Internal buds or gemmules are an assemblage of different types of cells, usually formed internally at the base of the sponges, that can be freed after extensive tissue damage and then ferried elsewhere to develop into an adult sponge. Finally, the pluripotent capacity of many sponge cell types may allow fragments of the sponge body accidentally separated from the parent as a result of wave impact during storms or the activity of predators, including humans, to regenerate a copy of the original organism.

However, even though sponges do not move, attached as they are to solid benthic surfaces, they do reproduce sexually. Sexual reproduction is defined as the formation of a new organism by the natural fertilization of an egg by a sperm cell.[9] Sponges can be either gonochoristic (unisexual) or hermaphroditic, oviparous or viviparous. Fertilization is internal in some species and external in others.[7, 8] Internal fertilization is made possible owing to the water column functioning as a sperm vector, and to the particle-filtering properties of those animals. Sperms can be released by a sponge and, when ferried to another specimen of the same species, participate in internal fertilization. It is remarkable that this mode of fertilization was even possible at the outset of animal life.

The larvae resulting from fertilization have some locomotive abilities through small hair-like body projections called cilia. However, at some point they will settle and stick to a solid support, where they will continue to develop to the adult stage.[7, 9] Likewise, *Thiomargarita* bacteria can attach themselves to gastropod shells and other benthic substrates. This implantation or nidation pattern shared by the gammaproteobacterium and sponge larvae seems to show the relationship that we are positing and most probably heralded a similar property of the mammalian embryo. The attached *Thiomargarita* cells will then elongate in stalk-like forms before budding off spherical daughter cells from their unattached poles,[10] a model that is very similar to sponge propagation by external budding. The cnidarians, which also have preserved one of the most ancient configurations of the animal lineages, share those properties as well.

Creation of Meiosis in Eukaryotes

How then did the organism resulting from the viral infection of a bacterium get equipped with meiosis, an asymmetric mode of cellular division? Perhaps, a brief overview of phage-bacteria interactions will shed some light on this question.

Bacteriophage virions are commonly released from their bacterial incubators by a lytic mechanism. In the usual lysogenic infection of a bacterium by a phage, the viral prophage integrates the bacterial host genome. The newly modified bacterium, called a lysogen, will continue to multiply by binary fission, and the combined proviral and bacterial genomes will be transmitted to the daughter cells (Fig. 2).

During this period, while the two genomes remain distinct, we believe that interactions involving lateral genetic transfer from

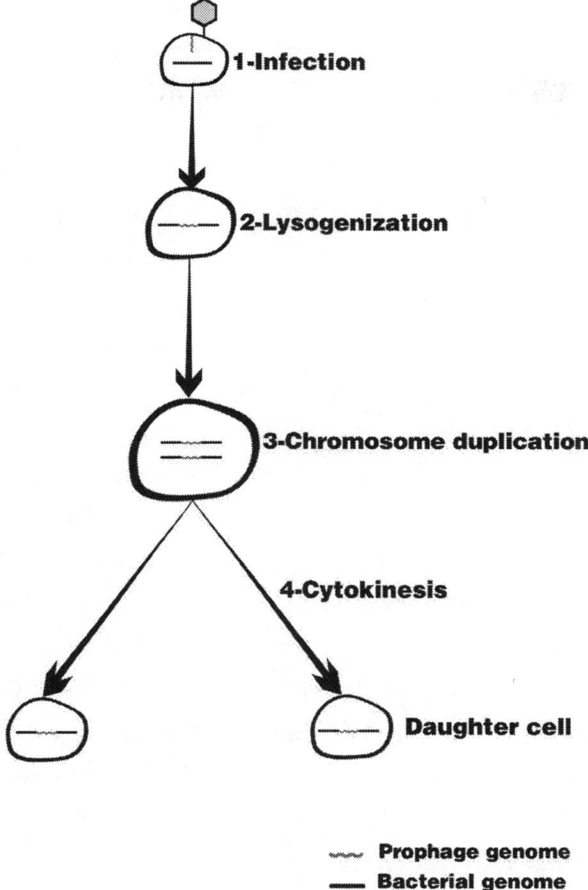

Figure 2: Formation of a lysogen.

the bacterial to the viral genome may have occurred. When, under specific circumstances, the prophage expression is induced, the cell will begin a lytic process made of the successive steps of prophage genome excision, followed by its replication, production of new virions, and lysis of the host. This scenario is a zero-sum game played by the bacteriophage, where it hits the jackpot (as it were) while the bacterium loses practically all or almost (except probably some of the bacterial genes that may have been laterally transferred to the phage genome). Unlike conventional viruses, however, retroviruses typically do not lyse their hosts as an exit strategy but instead bud from cells.

On that basis, we posit that in the process of gametogenesis out of the somatic cells allotted for reproduction in the first multicellular eukaryotes, the prophage-derived DNA will excise itself from the hybrid genome and, in a sort of asymmetric cellular division, be expelled from the parent cell. This pattern is quite different from that of bacterial lysogens, where induction of prophage DNA is followed by its excision, replication, and exit by cellular lysis. The forms of expulsion of genetic material from the eukaryotic cells will vary. For instance, it may be a polar haploid nucleus in some cases, or within a miniature cell called a polar body in others (see Chapter 3 for more on this). In the latter, the process will be one of unequal cytokinesis, analogous to oogenesis in the familiar XY sex-determination system. The viral-derived DNA will segregate and bud as a polar body, leaving in the remaining cell (now called the secondary oocyte) the genetic material of bacterial origin, notwithstanding the requirement of meiotic recombination that has taken place during that process. Again, we make this contention of viral-derived DNA exit considering the usual phage-bacteria interactions.

Eukaryotic meiosis is then established. This scenario, preceded by cellular DNA replication, is a win-win strategy for both protagonists of eukaryogenesis. The two cells will complete a classic meiotic division that will lead to the formation of the very first gametes. The cell harboring the retrophage-derived genome or first polar body will eventually differentiate into haploid sperm cells; the other, containing the bacterial-derived genome will, in the XY sex-determination system, for instance, differentiate into a diploid zygote and a second polar body at the moment of fertilization (see Chapter 3 and Figure 5). Theoretically, were the second polar body to form a sperm cell, it would genetically be of bacterial origin. It can therefore also function as a sperm, the haploid nucleus, or the polar fragment that results from the asymmetric division of a eukaryotic cell. In short, progenitor cells that underwent XY-type meiosis could produce three different types of gametes: eggs with a bacterial genome, sperms with a viral genome and sperms with a bacterial genome. A more detailed analysis of ancestral gametogenesis will be performed in Chapter 3.

The first gamete-producing eukaryotes were then hermaphrodites, able to produce both sperms and eggs. This is, however, an argument about the possible outcomes of one cell undergoing meiosis in the hydrosphere of the early Earth. It is unclear, however, whether in the real world of early meiosis, XY organisms used all three types of gametes. It is possible that some individual organisms specialized in the production of specific functional gametes.

Fertilization as encountered today for any species started with the lysogenic "infection" of a bacterium by a bacteriophage. One phenomenon (viral infection) is the mirror image of the other (fertilization or fecundation). Lysis in the phage-bacterium realm is mirrored, for instance, in the world of eukaryotes by the momentous event when a chick pecks at an egg wall to hatch, or when a woman's amniotic sac ruptures during labor; only that in those instances, the newborn is assuredly the combined product of both progenitors rather than of one exclusively, and the ovum has played the role of more than just an incubator.

Gametes and organisms of all species are vehicles used by the genomes under the stewardship of the non-protein-coding elements to ensure their reproduction and perpetuation in time. Spawning and copulation represent, in this context, mechanisms used by metazoans to convene for fertilization the products of meiosis—the sperm and the oocyte. Figure 3 shows a diagram illustrating the origin and modes of natural animal fertilization, including copulation.

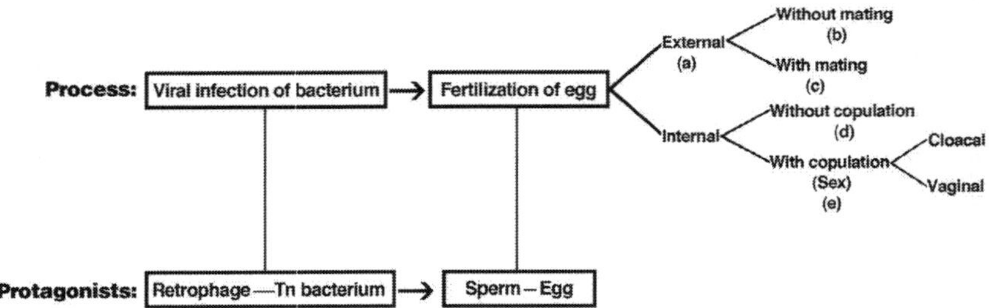

Figure 3: Diagram of the origin of sex and animal fertilization (human contributes to the methods of animal fertilization with Assisted Reproductive Technologies [ART]).

a) Exclusively in aquatic milieux (spawning).

b) Example: one strategy of sexual reproduction in sponges, cnidarians.[7]

c) Example: Most anuran species.[11, 12] The male frog or toad hops onto the back of a female, typically attracted to it by its vocal display. It then covers with sperm the eggs that its partner has just spawned.

d) -Self-fertilization or selfing: When fertilization involves the fusion of gametes from a single individual, as is often the case in the nematode *Caenorhabditis elegans*.[13-16]

   -In insects such as Collembola (*Sminthurus viridis*), the male deposits a bag of sperms (spermatophore) on a surface that a passing female may pick up with her genital opening.[17]

e) Aquatic mammals and most land animals.

Cellular exit from meiosis and the egg-to-embryo, meiosis-to-mitosis transition will be accomplished only after fertilization, owing to the sperm-triggered increase in intracellular free Ca^{2+}.[18] The fusion of the gamete cells induces the increase of this cation inside the egg. The sudden surge of free calcium ions sets off several intracellular processes, one of which is the fusion of the male and female pronuclei before the first mitotic division that generates the blastomeres of the two-cell embryo. We suggest here that this process is the result of the expression of genes previously expelled from a cell, as is the case in certain enveloped viruses or any polar body. This is indeed the case of the origin of the proteins syncytin-1 and syncytin-2, known mediators of the fusion of human placental trophoblast cells. The proteins syncytin-1 and syncytin-2 are encoded respectively by the envelope gene of the human endogenous retroviruses HERV-W and HERV-FRD.[19-21]

In fact, this calcium-mediated fusion of sperm and egg pronuclei bears a striking resemblance to the cell fusion initiated by certain viruses. In 1965, Okada and Murayama demonstrated that the Sendai virus-induced fusion of Ehrlich ascites cells required Ca^{2+}.[22] The Sendai virus (SeV) is also referred to as the hemagglutinating virus of Japan (HVJ).[23] The Okada experiment was again confirmed in 1982 when Hallett and collaborators demonstrated that SeV induces a rise in intracellular free Ca^{2+} before cell fusion.[24] More recently, calcium ions were found to be a requirement for cell fusion and syncytia formation by the respiratory syncytial virus (RSV) and HIV-1.[25, 26] Multicellularity as it is known in eukaryotes was built by viruses.

The Universal *Bauplan* (Body Plan)

According to the bilateral descent theory, eukaryotes emerged as a novel domain of life when some bacteriophage viruses fused their genetic material with that of their hosts. Their cellular properties will be largely derived from their bacterial ancestors. To the eukaryotic cell, however, the viral side will contribute, among other structures, intracytoplasmic organelles such as the nucleus, mitochondria, and chloroplasts. Eukaryotic multicellularity will organize the cells into tissues, organs, and systems. We propose here that the general body plan of organisms is also derived from their bacteriophage ancestors.

Indeed, we have seen that the mammalian sperm cells display morphological similarities with the Caudovirales. The latter exhibit a tadpole-like craniocaudal axis[27] and there is no reason to consider that either their retrophage ancestors or those that contributed to the emergence of the animal kingdom were designed differently. That head-tail/flagellum structure is generally represented in the sperm cells of animals and most certainly in those of mammals. We propose that this viral design is at the origin of the animal antero-posterior axis, a factor of the bilaterality expressed without ambiguity in almost all the species of this kingdom.[28] Bilaterian animals are said to display bilateral symmetry that results from the right-angled (orthogonal) intersection of a craniocaudal or antero-posterior (A-P) body axis with a dorso-ventral (D-V) axis.[28] Those two axes of polarity are then derived along a single body plane that divides the body into a left and right side.[29] The group of genes responsible for the embryonic segmental patterning of the A-P axis in bilaterians are called Hox genes, while the decapentaplegic (dpp) gene is implicated in the control of the dorsal-ventral specification.[28, 30, 31]

Contrariwise, Cnidarians such as jellyfish, sea anemones, and corals are generally believed to display radial symmetry. However, some Hox genes were found to be expressed in a staggered fashion over the oral-aboral axis of *Nematostella vectensis*, a sea anemone, which also exhibits an asymmetric dpp expression along the dorsal-ventral or secondary body axis.[28, 32, 33] Therefore, vestiges of our Caudovirales origin can be found in the genes and body plan of all animals. Suffice to say that the Hox genes, at least, obviously originated from the bacteriophage ancestors of the animal kingdom. As a matter of fact, the helical tail of the Caudovirales is also made of segments or "stacked disks,"[34] a plausible feature of the retrophage ancestors that is, we submit, at the origin of the neurologic segmentation of animals. The genetic material, tightly folded in the head of the Caudovirales and which proceeds into the tail, is replaced in our *Bauplan* with a central nervous system similarly compacted within the skull, and that continues in our spinal cord, our genome having been moved to within our cells. Our human body plan roughly mirrors the profile of some tailed viruses, especially the outline of bacteriophage T4 (Fig. 4).

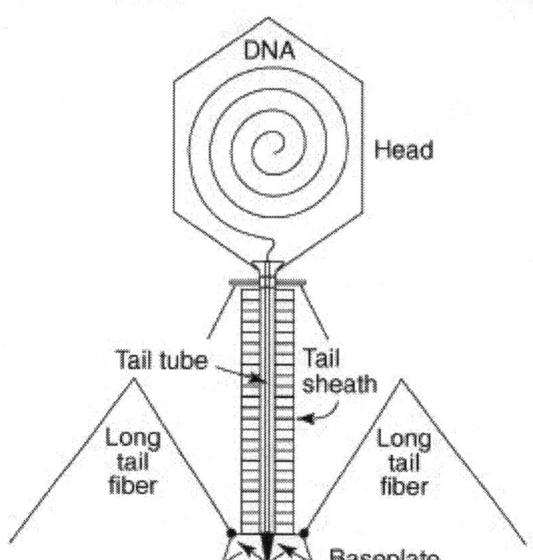

Figure 4: Schematic representation of the structure of bacteriophage T4.

(Adapted from Leiman, P.G., Arisaka, F., van Raaij, M.J. *et al.* "Morphogenesis of the T4 tail and tail fibers," *Virology Journal*, 7:355 [2010], https://doi.org/10.1186/1743-422X-7-355. Reprinted through the use of their Creative Commons Attribution License.)

CHAPTER

3

On Sexual Differentiation: Or, How Eve Begot Adam, All by Herself!

> Then that little man in black there, he says women can't have as much rights as men, 'cause Christ wasn't a woman! Where did your Christ come from? Where did your Christ come from? From God and a woman! Man had nothing to do with Him.
>
> —Sojourner Truth. "Ain't I a Woman?" speech.

A novel organism emerges from the fusion of the genetic material of a virus with a bacterium. With this interaction was born the new domain Eukaryota, and with each of the different strains of virus pairing with bacteria, a specific lineage was created. Members of a lineage, however, were not sexually differentiated. The first metazoans were hermaphrodites (male/female entities),[1] i.e., a given organism had the capacity to produce both sperms and eggs. Hermaphrodites are essentially somatic females who use the polar bodies of their oogenesis as sperms to potentially either self-reproduce (selfing) or to outcross.[2-4] The essential characteristic of the female soma is indeed the potential to produce eggs. We believe that outcrossing would have been favored over selfing because of the biological disadvantages of inbreeding.[5-8] Hermaphrodites may use varied mechanisms to prevent self-fertilization. For one, in the hermaphroditic sponges, the production of sperms and eggs is asynchronous.[9] Furthermore, viviparous sponges specifically keep their oocytes for internal fertilization and embryo incubation but entrust their sperms to the water currents to be captured by other conspecific sponges. This last property has earned them the qualifier of spermcasters.[10]

Many investigators also suggest or support the theory that hermaphroditism was the ancestral sexual condition from which eventually evolved the separation of the sexes between distinct individuals.[1, 11-13]

Prior to sexual differentiation, the chromosomes of the ancestral species were arranged exclusively into autologous pairs or autosomes. However, a pair of XX and ZZ autosomes, in the respective ancestors of mammals and birds for instance, eventually differentiated into the genetic XY and ZW sex-determination systems. Haplodiploidy, on the other hand, constitutes the sex-determination framework of the entire insect order Hymenoptera (ants, bees and wasps). In this system, females are diploid, and fertile males are haploid. Haplodiploidy stems originally from an ancestral *feminizer* (*fem*) gene represented equally on a pair of homologous chromosomes in the ancestral organism.[14, 15]

With sexual differentiation, each type of gamete will be produced by separate individuals (unisexualism in animals, dioecy in plants). However, before we delve here into the mechanisms of sexual differentiation, it is useful to consider the sexual reproductive strategies of the hermaphroditic ancestors whose chromosomes were all autosomes. The essential characteristic of the genome of the ancestral hermaphrodite is its viro-bacterial signature. The main task of such an organism in sexual reproduction will be to ensure the maintenance of this pattern of genetic heritage in the diploid cells of its descendants.

Oogenesis and Fertilization in the Hermaphrodite Ancestors

As previously mentioned, diploidy in ancestral hermaphrodites is achieved entirely by pairs of autosomal chromosomes, one of which is of viral and the other of bacterial origin. In the XY and ZW sex-determination systems, the second meiotic division of oogenesis with the expulsion of the second polar body is not completed until after fertilization,[16-19] a strategy we believe is inherited from the ancestral organisms. Thus, self-fertilization in XX ancestors could have occurred only, if permitted, between the oocyte and a sperm derived from the first meiotic division, i.e., of viral origin. The proposed steps of oogenesis in the first XX eukaryotes are roughly schematized in Figure 5.

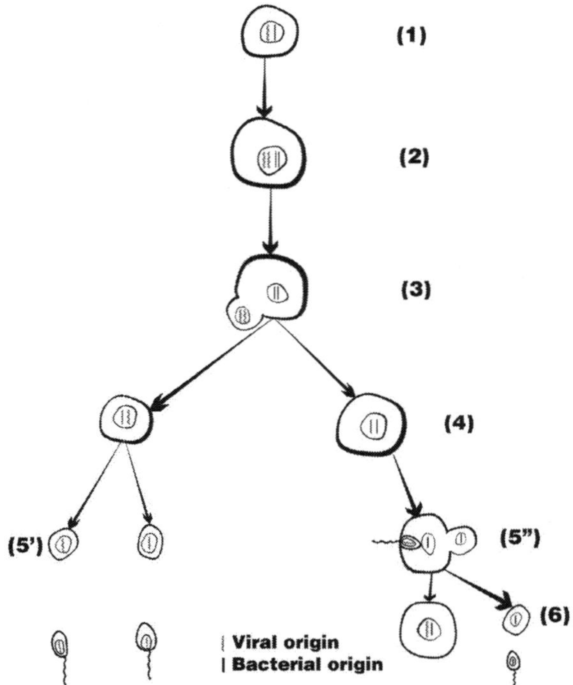

Figure 5: Oogenesis and fertilization in the hermaphrodite XX ancestor.

1- Somatic pre-meiotic cell (2n) destined to become germ cell after sexual differentiation. For the sake of clarity, only one pair of autosomes is represented: one homolog of viral (wavy grey stick), the other of bacterial (straight black stick) origin.

2- Primary oocyte: pre-meiotic chromosomal replication (4n).

3- Formation of the first polar body for the expulsion of the replicated viral homolog from the cell.

4- Meiosis I: first cytokinesis with the formation of the secondary oocyte and the first polar body.

5'- The first polar body divides (meiosis II) and produces two daughter polar bodies that can play the role of sperms.

5"- Fertilization of the secondary oocyte in progress.

6- Completion of meiosis II for the secondary oocyte after fertilization has occurred. The outcome is a fertilized egg and expulsion of the second polar body.

Likewise, the ancestral state that led to the haplodiploidy sex determination system is considered to have derived from autosomal pairs carrying the *feminizer* (*fem*) gene. A major feature of this system is that meiotic divisions of oogenesis are not cellular, but rather limited to the nucleus. The divisions of the cell's nucleus result ultimately in the formation of four aligned haploid daughter nuclei. Those four nuclei are usually arranged in a row parallel to the longitudinal axis of the egg and in a periplasmic location.[20] One of them, whose fate is to become the egg pronucleus, is larger than the other three and positioned in a central location. As for the oocyte itself, it does not divide during meiosis; instead, it continues to accumulate yolk and to increase in size.[20, 21] Three of the four intracellular products of meiosis are usually lost to reproduction as polar nuclei and will degenerate. The fate of the remaining pronucleus is to either develop as a haploid male if it remains unfertilized or to produce a diploid zygote if it fuses with a sperm nucleus.

However, in the thelytokous reproductive modality adopted by queenless worker bees of *Apis mellifera capensis*, which we propose as a possible fertilization strategy of the ancestral hymenopteran, the diploid zygote derives from the fusion of the egg pronucleus with one of the polar nuclei acting as sperm. *Apis mellifera capensis* (hereafter sometimes *Capensis*) is a subspecies of honeybee in South Africa. The most telling aspect of this strategy is that this fusion is not random; it happens generally between the egg pronucleus and the central descendant of the first polar body (central fusion). The other polar bodies will eventually degenerate while the zygote nucleus moves to the center of the egg. The zygote will then divide mitotically to form the embryo, a female in the case of *Capensis*,[20, 22] but a hermaphrodite, we submit, in the ancestral hymenopteran. An alternative reproductive strategy that the ancestral hymenopteran could have used is the formation of embryos by ameiotic parthenogenesis, as has been reported in *Capensis*.[23, 24]

The oogenesis diagram in the haplodiploidy system bears some similarities with the X- or Z-based systems, with some minor differences. One of them is that, in the former, meiosis II moves along without the need for prior fertilization. Referring to several sources, [20, 21, 24] we have illustrated the steps of oogenesis and fertilization in the ancestral hymenopteran as shown in Figure 6.

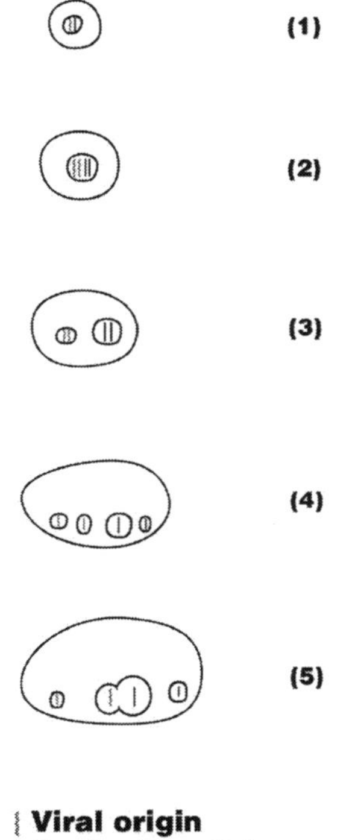

Figure 6: Oogenesis and fertilization in the hermaphrodite honeybee ancestor.

1- Somatic pre-meiotic cell within which is represented the pair of autosomes (2n) destined to become sexual chromosomes: the chromosome homologs of viral and bacterial origin are represented respectively as wavy grey and straight black sticks.

2- Primary oocyte: pre-meiotic chromosomal replication (4n).

3- First meiotic division. It is a reductive nuclear division that produces two nuclei: one larger (egg pronucleus) than the other (first polar nucleus).

4- Second meiotic division, again reductive for the egg pronucleus, leading to the formation of four haploid nuclei (the larger one—the egg pronucleus—in a central position).

5- Fertilization in the hermaphrodite hymenopteran is posited to have happened by the fusion of the two central nuclei (central fusion) as in *Apis mellifera capensis*.

We contend that in all the three modalities described—the XX, the ZZ, and the haplodiploidy systems—the ancestral organisms "self-reproduced" as hermaphrodites through the fecundation of the oocyte by a polar body or nucleus acting as sperm in a way capable of maintaining the viro-bacterial nature of the genome. In other words, an oocyte with a bacterial-derived genome is fertilized by a sperm with a viral-derived DNA, or vice-versa.* To this end, ameiotic parthenogenesis, as in the case of *Apis mellifera capensis*, could have also been used as a reproductive modality.

Modalities of Sexual Differentiation

Hermaphroditism, the ancestral reproductive strategy, is still extremely well documented among many species. It is estimated to occur in roughly 94% of plants and, excluding insects, about a third of animal species.[4, 13] For instance, as previously mentioned, it is the case in most sponges. The female nematode *C. elegans* can facultatively use her polar bodies as sperm to self-fertilize or, mate with a male that has evolved according to mechanisms hereafter described. The garden snail (*Cornu aspersum*) reproduces mostly by mating and exchanging sperms with another snail.[25] Furthermore, the predominant group of land plants, which are called angiosperms or flowering plants, is mostly co-sexual (hermaphroditic or monoecious).[1, 11-13] As previously argued, this reproductive choice is a vestige of ancient biology, rather than a novel method that has evolved from unisexual reproduction. The extant hermaphrodites of the animal kingdom are female organisms that can produce sperm, and some of which can be self-fertile (parthenogenetic reproduction).[4]

* To our knowledge, the details of oogenesis in the ZW system where the female is heterogametic have not yet been elucidated.

However, while it is true that most plants have remained hermaphroditic or monoecious, in the animal kingdom unisexualism has prevailed over evolutionary time. Most animal ancestors will evolve into exclusive unisexual (male or female) organisms as a result of different sex determination systems. In other words, any one individual will produce only one type of gamete. Reconstitution of diploidy will then involve the rendezvous of gametes from two different individuals. Sex or copulation, one of the different mechanisms that have emerged for this purpose, is a natural attempt of mixing one's genetic information with that of another individual with a goal (unconscious, except sometimes in humans) of it to be carried by an organism of a new generation. Sexual reproduction is both a risky and a "costly" pursuit.[26] For instance, male animals in many species must give up opportunities to find food, expose themselves to predation during a mating venture, or risk bodily harm or even death at the "hands" of rivals, while trying to gather a harem of females. Pregnancy and delivery are not either without morbidity or mortality risks for the female. However, as we have shown, sex is woven into the fabric of eukaryotic life, the eukaryotes having been brought up from essentially the "fecundation" or "eukaryotic fertilization" of bacteria by viruses. All measures thus taken to prevent "reproduction" constitute in a way a co-optation of sex. In some higher organisms, such as humans, bonobos, or dolphins, sex can also be entirely recreational and play the emergent function of a social tool in addition to its reproductive role. In humans, we venture, the continual provision of sex helps to keep couples together even after the female reproductive age, in part so they can continue to raise children and contribute to the rearing of grandchildren.

A wholesale discussion of gender differentiation is, however, beyond the scope and goal of this book. But we cannot resist wondering how the process of producing sperms and eggs, ancestrally confined within one organism, evolved to be divided between separate individuals.

In addition to genetic factors, sex can be determined by environmental cues, such as temperature.[27] We will, however, only consider here the genetic sex-determination pathway, most particularly in animals. The pioneer eukaryotic cells contained two sets of chromosomes, of which one was of viral, the other of bacterial origin. The organisms, having not yet differentiated sexually, contained only autosomal chromosomes within their cells. Heteromorphic sex chromosomes such as the X and Y of mammals, or the Z and W of the avian model, are thought to have evolved from a pair of ancestral homologous autosomes.[1, 13, 28, 29] The two autologous chromosomes evolved away from each other by several mechanisms that include cessation of recombination, specialized gene content, and accumulation of noncoding sequences from transposable elements in the Y and W chromosomes, for instance.[30, 31] While hermaphroditism in the ancestral species was maintained by the bilateral or viro-bacterial origin of their chromosomes, unisexualism in their descendants will derive from certain combinations of alleles or sexual chromosomes.

Research done in the hermaphroditic *C. elegans* and other species has in fact revealed the existence of autosomal genes that can be mutated in a way able to transform the sexual phenotype of their carriers.[2, 32] While across the board sex chromosomes evolved from autosomal pairs, the details of that evolution vary among the species. Out of the different genetic sex determination systems, we will here expand briefly on the three most common modalities.

Indeed, in eukaryotic organisms, the two sexes can be differentiated by their sexual chromosome representation. The XY sex-determination system is found in mammals and several other species, including insects such as *Drosophila*, plants such as papaya (*Carica papaya*), and the dioecious plants of the genus *Spinacia*.[1, 12] Basically, in the XY system, females produce only one type of gamete

(X) and are then said to be homogametic, while the heterogametic males produce both X and Y sperms. Females are diploid XX and males are XY, or XO as in the case of the male *C. elegans*.

When it comes to the fly *Drosophila melanogaster*, sex is determined by the X:A ratio or the number of X chromosomes relative to the number of autosomes.[33, 34] This model represents the commonly accepted theory of sex allocation in the fruit fly, notwithstanding the alternative version that the X chromosome dose instead of the X:A ratio signals sexual fate in *Drosophila*.[35, 36] In females, with two X chromosomes (XX), this X:A ratio is 2/2, while in males, which carry only one X (XY), the ratio is 1:2.[33] The pattern is not much different in nematodes, which do not carry a Y chromosome. The self-fertile *C. elegans* has a XX sexual genotype, while the male is XO. Most related nematode species also have an XX female/XO male system.[2]

The roles are inverse in the ZW system, where heterogametic ZW organisms are females while the homogametic sex is male (ZZ).[31] The ZW system mediates sex determination in birds, reptiles, and most insects. Details of oogenesis in the ZW system, particularly how two types of oocytes could result from oogenesis, have not yet been clarified comprehensively. The female sex in the XY system is homogametic, in the sense that oogenesis leads to one type of secondary oocyte (X). This model is not valid for the ZW system, where the heterogametic female will produce two types of oocytes, Z and W. However, no data is available specifically about the details of this process in the ZW system.

The haplodiploidy system, in turn, determines the sex in the insect order Hymenoptera and sporadically in some other animals, such as bark beetles (*Scolytidae*), ticks, and mites (*Acarina*), etc.[37] It is found in about 20% of the animal kingdom. In honeybees (*Apis mellifera*), sex determination is governed by heterozygosity at the multiallelic complementary sex determiner (*csd*) gene. Female bees develop from fertilized eggs and are heterozygous at *csd*. Bees that are either homozygous or hemizygous (haploid) at *csd* are males. Haploid males normally develop from unfertilized eggs and are fertile. Diploid or homozygous males are sterile, or in the rare species where they are fertile, they produce triploid sterile offspring.[14, 37] Diploid male honeybees are usually eaten by adult female workers.[15, 37]

Hasselman and Beye have demonstrated that while there is a relative increased recombination rate in the region surrounding the sex determination locus (SDL) that contains the *csd* allele, no recombination was found within the locus itself,[38] a finding on par with the cessation of recombination between the sex chromosomes when they form a heteromorphic pair in the respective XY and ZW sex-determination systems. The gene *csd* itself arose from the relatively recent duplication of the progenitor *feminizer* (*fem*) gene. The hymenopteran *fem* gene and the Drosophila principal sex-determining gene *transformer* (*tra*) seem to have a common evolutionary origin based on a few similar characteristics, namely a conserved sequence motif, the arrangement of certain domains enriched with identical amino acids in the gene product, and their equivalent function in regulating sex determination.[14] The *fem gene* is ancestrally conserved, while its duplicate, *csd*, formed several distinct alleles. Starting from a primary signal in both the honeybee (*fem* gene) and the fruit fly (*tra* gene), a network of gene interactions—much more complex in the latter—initiate a cascade of regulatory (downstream) responses, including alternative splicing of primary transcripts, the action of stop codons, etc., that finally implement two different routes of sexual differentiation.[14,15, 33]

The general trend of sexual differentiation is the evolutionary divergence of certain ancestral genes or autosomal pairs. For instance, the *csd* gene emerged from the duplication of a conserved progenitor *fem* gene and then allowed heterozygosity by its multiallelic character. The Y chromosome of the XY system results from the degeneration of an ancestral X and evolved the *SRY* gene, which acts as a signal to establish maleness. Female is the default sex, established by the absence of the *SRY* gene. A strikingly common feature of the Y and W chromosomes across taxonomic groupings is the accumulation of noncoding sequences or of transposable elements, which is referred to as the Y and W degeneration, thought to be due to a lack of genetic recombination.[31] Thereby, in the ZW system, the W chromosome evolved from the degeneration of an ancestral Z, during which time it lost most of its original genes.

The Z-linked DMRT1 gene has subsequently evolved to establish maleness in the homogametic ZZ.[27] According to the Z dosage hypothesis, a lower dosage of DMRT1 in the female ZW would prevent testicular differentiation as compared to the ZZ male.[27] Nonetheless, the silkworm *Bombyx mori* has a W-linked feminizing gene, called *Feminizer* (*Fem*), that establishes femaleness, with male being the default state.[31, 39]

It is worth noting that it is through alternative splicing of the primary mRNA transcript of a single respective gene that the two different routes of sexual differentiation are adopted in both *D. melanogaster* and the honeybee. Alternative splicing is a processing modality of the mRNA transcript that allows for the production, from a single gene, of more than one isoform of a protein. It entails that, based on different signals, certain exons in one case may be processed as introns in another one or vice versa. This diverging mechanism in the case of the *tra* gene of *D. melanogaster* or the *fem* gene of the honeybee allows for the production of a male- or female-specific protein isoform based on a chromosomal or genetic primary signal. In the hymenopteran, this response is governed by the allelic composition of the genome at the single-locus *csd* gene: Heterozygosity leads to a female phenotype whereas haploidy or homozygosity ensures a male phenotype. In *Apis mellifera capensis*, for instance, assuming the absence of meiotic *csd* allele recombination as previously reported,[38] the fusion of two haploid nuclei each derived from a different product of the first meiotic division will result in female offspring, while the outcome of the fusion of nuclei derived from the same product of meiosis I (terminal fusion) will be diploid male offspring.[24] In the fruit fly, the primary genetic signal of the binary switch of the sex determination pathway is provided by the X:A ratio as previously explained.

CHAPTER

4

The Roots of Biodiversity

Diversity is one of the most striking characteristics of life. The living world encompasses phenotypes from the simplest viroids to the most complex mammals. Making sense of this vast repertoire of life is no small task, and that is where taxonomic classification has shown its usefulness in biology, by helping to distinguish and categorize the diverse forms of the living. The 18th-century Swedish naturalist Carl von Linne devised the systematic classification of cell-based life that is still in use today. According to this system, cell-based life is classified around a fundamental unit of taxonomy: the species. Each species is named according to a binomial (two-part) nomenclature, comprising the capitalized genus followed by an adjective or an epithet: for example, *Homo sapiens* for humans or *Ursus maritimus* for the polar bear. The physical contrast among a great number of eukaryotic species is generally appreciable to the human senses. How nature has produced species as different as the horse (*Equus caballus*), the apple tree (*Malus domestica*), or the house mouse (*Mus musculus*), for instance, is puzzling. It is with good reason that the question of the origins of this diversity has occupied the mind of humanity at least since Aristotle. In the Western world, this interrogation knew a certain fervor in the 19th century with leading figures such as Lamarck, Cuvier, etc., and culminated with the publication by Charles Darwin in 1859 of his magnum opus, *On the Origin of Species.*

The existence of many definitions of the concept "species" betrays some difficulty to encapsulate it and may be a hint to the continuity of nature. We take note here of the sentiments of the paleontologists Niles Eldredge and Stephen Jay Gould that the concept "species" fails to appropriately render the continuum or the unbroken bond of life as studied longitudinally.[1] They rightly point out that the Linnean world of discrete and static entities apprehended in a snapshot of time only intersects with their own, in which they study the fate of lineages over relatively vast expanses of time. In their own words: "The hierarchical system of Linnaeus was established for his world: a world of discrete entities. It works for the living biota because most species are discrete at any moment in time. It has no objective application to evolving continua . . . Linnaeus would not have set up the same system for our world."[1]

But under all its interpretations, the notion of species is rendered as the basic category of cellular life. While these definitions*[2] are separate and exclusive, each one of them, we surmise, may contribute something of value to the full measure of the concept. For the biological species concept, the key defining characteristics of sexually reproducing conspecific organisms are their capacity to interbreed and to produce fertile offspring. They are then said to be reproductively isolated from heterospecific populations. Obviously, this definition would not include a single-celled organism such as *Amoeba proteus* that reproduces by binary fission or mitosis.

It is safe to say that eukaryotic life is quite diverse. Camilo Mora and colleagues estimated the number of eukaryotic species at around 8.7 million.[3] However, to date, only 1.8 million species have been named.[2] Nonetheless, the enigmatic factor that accounts for all this eukaryotic biodiversity has puzzled thinkers, again since the days of the ancient Greeks. Darwin called the process by which

---

* Biological (most commonly used), morphological, phylogenetic.

the individual species were formed "the mystery of mysteries." His theory of the origin of species by means of natural selection laid out in *On the Origin of Species* did not satisfy all the questions of some, nor converged with the answers of others.

The prevalent understanding of the origin of the species surmises at its core that "descent with modification" or random germline DNA mutations engineered the apparition of new species from a common ancestor. The transitional forms and the casualties of natural selection would explain the "extinct species" of the catalog of life. Extant species would have been endowed with higher fitness-conferring germline DNA mutations compared with the extinct ones. But overall, all species would have resulted from a common eukaryotic ancestor by successive, random DNA mutations. The genome differences between any two species would result from those said mutations and are expressed as differences in the bases A, C, G, and T that form the rungs, if you will, of the DNA double helix. This is a rough summary of the neo-Darwinian worldview.

Indeed, comparative genomic analysis has revealed relatively small differences between some species with very different phenotypes. We humans have a substantial genetic homology with mice, and share over 98% of our genome with the chimpanzee, our closest species.[4] The human genome, for instance, is made of about 3 billion base pairs, while the average number of DNA differences between humans and the common chimpanzee (*Pan troglodytes*) is 1.23 for every 100 bases of DNA.[5] Briefly put, according to neo-Darwinism (the current paradigm in biology), a two-prong mechanism—undirected DNA mutation and natural selection—accounts for the origin of the species from a common eukaryotic ancestor. This mechanism infers the survival of some species and the extinction of others. Darwin himself characterized the topic of species extinction as "being intimately connected with natural selection."[6] Let us take a closer look at DNA mutations.

DNA Mutations and Speciation

A DNA mutation is a change in the nucleotide sequence of the chromosome. Genetic mutations are not infrequent phenomena in cellular organisms. A typical mammalian cell is estimated to sustain DNA damages at a rate of over 100,000 molecular insults per day.[7] Drake and colleagues have estimated that each human zygote accumulates 64 new mutations.[8] Furthermore, as DNA is replicated before each cell division, the sheer number of those events in multicellular organisms over time affords on each occasion a new opportunity for mutations.

A gene mutation can either be hereditary—that is, transmitted from the previous generation to the next through the germline cells—or acquired during the lifetime of an organism. Acquired or somatic mutations occur in non-germline cells. The mutations relevant to the neo-Darwinian speciation process are those transmitted to the next generation through the germline cells.

In most cases, the cell will detect and repair the mutation, or it may be given a command to commit suicide. In this latter case, the process is called apoptosis or "programmed cell death." Despite the many DNA repair tools available to the cell, such as the proofreading mechanisms of DNA-dependent polymerases—the enzymes that replicate DNA from a DNA template—some replication errors may slip through. A DNA mutation that has managed to survive all the cellular repair mechanisms can have either a neutral or a non-neutral effect. A neutral or silent DNA mutation can happen in many ways: the nucleotide switch either does not change the amino acid or, if it does, the corresponding protein remains functionally identical. Nucleotide changes that do not affect the amino acid added to the protein sequence are called synonymous mutations, as opposed to nonsynonymous. Indeed, because of the redundancy of the genetic code, more than one codon may specify a particular amino acid.

Known beneficial DNA mutations are extremely rare, to say the least. Radiation-induced mutation experiments on the fruit fly *D. melanogaster* were performed in the early part of the 20th century by pioneers such as H.J. Muller at the University of Texas and Thomas Hunt Morgan at Columbia University. Muller reported that the lethal mutations "greatly outnumbered" the nonlethal ones that produced a visible morphological abnormality.[9] Sterility, which constituted one more type of deleterious effect in the flies he treated, was not initially recognized. On another note, Muller added that, in addition to causing some genes to mutate, the X-ray treatments can change their place on the chromosome.[9]

The work started by those pioneers was pursued by other well-known names in biology, such as Theodosius Dobzhansky whose radiation treatments failed to create any new lineages or confer any significant advantages to the insects subjected to the procedure. Instead, these experiments inflicted all kinds of disabling deformities on the surviving flies. Mutants without eyes, others with short or twisted wings, and yet others with legs growing out of their heads where antennae should be (the Antennapedia mutation) accounted for only a sample of those debilitating malformations. Concerning the claim that DNA mutations constitute the raw materials of evolutionary changes, Dobzhansky, in his book *Evolution, Genetics, and Man,* lamented the many deleterious effects of mutations in Drosophila melanogaster.[10] Many flies even die as a result. Here is how he described some of the effects of radiation-induced gene mutations: "Most mutants which arise in any organism are more or less disadvantageous to their possessors. The classical mutants obtained in Drosophila usually show deterioration, breakdown, or disappearance of some organs . . . Many mutants are, in fact, lethal to their possessors. Mutants which equal the normal fly in vigor are a minority and mutants that would make a major improvement of the normal organization in the normal environment are unknown."[10]

Nevertheless, as a founding father of neo-Darwinism, he quickly added that DNA mutations such as the ones that confer antibiotic resistance to bacteria, or DDT resistance to the houseflies, are useful. We will now briefly summarize some cases of purported positive or useful DNA mutations.

Lactase Persistence Mutation

One oft-mentioned beneficial human DNA mutation is the one responsible for lactase persistence in adults in some societies. In humans, as in all mammals, the expression of the lactose-digesting enzyme or lactase normally decreases after weaning. However, a point DNA mutation near to the lactase gene located on chromosome 2 has allowed for the persistence of this disaccharidase in some human populations. This mutation is inherited in an autosomal dominant manner. It is found mostly in populations in Western Europe, the Middle East, and in the African Masai and Tutsi.[11, 12] This situation is known as lactase persistence, per opposition to the lactase non-persistence of most sub-Saharans, East Asians, and the native peoples of the Americas and Pacific islands.[13] This beneficial mutation obviously provides its carriers with the ability to digest fresh milk, a source of energy that others cannot tolerate. However, upon closer scrutiny, one realizes that this advantage is not even the effect of a structural gene mutation. No change happened in the lactase gene (LCT) or the protein (lactase) produced. The lactase gene is not touched. Lactase persistence (LP) is the result of a regulatory change to the lactase gene, allowing the prolongation of its expression long after weaning, which is different from the normal mammalian condition. A few single nucleotide changes called single nucleotide polymorphisms (SNPs) have been associated with the lactase persistence (LP) trait in different populations. The first SNP identified was in northern European populations and located within an intron of a gene in sis position (upstream) relative to the lactase gene itself.[14, 15]

Xylitol Use by Aerobacter Aerogenes Bacteria[16-19]

The bacteria *Aerobacter aerogenes* strain PRL-R3, currently named *Klebsiella pneumoniae var. oxytoca* strain PRL-R3, is known to possess the enzymatic pathways necessary to metabolize ribitol and R-arabitol. These two five-carbon sugars belong to the pentitol family. The ribitol catabolic pathway was shown to consist of a ribitol dehydrogenase and a kinase enzyme to phosphorylate, with the help of ATP, the metabolite that has resulted from the action of the dehydrogenase. Subsequent studies have shown that those enzymes were inducible, which means that in the absence of ribitol in the growth medium, they were only synthesized in extremely low amounts by the bacteria; but once the substrate was introduced into the milieu, the bacteria responded by substantially increasing their production. Those enzymes are part of an operon system, i.e., their synthesis is under the control of adjacent regulatory genes activated only when the specific inducer—the substrate—is present. Careful observations have shown, however, that the wild type of the bacteria could not normally metabolize xylitol, another pentitol. Nonetheless, when enough of the bacteria (about 10^7) were inoculated into a medium containing xylitol, there was satisfactory growth from the cohort of bacteria that has gained the ability to degrade the pentitol through spontaneous mutation. It was puzzling to know how, on such short notice, the bacteria could mount the synthesis of a new set of enzymes and new metabolic pathways to degrade a new source of food.

That is when it was found that the bacteria did not require the synthesis of any new enzymatic activity to grow on xylitol. Because of the structural similarity between ribitol and xylitol, the dehydrogenase that evolved for the ribitol pathway could oxidize xylitol to an initial metabolite on which a specific kinase will then act. To grow on xylitol, the cells only required a regulatory mutation permitting the ribitol dehydrogenase to be synthesized continuously, rather than only upon induction by the presence of ribitol in the growth medium. Growth on the new substrate, xylitol, was made possible through a mutation that caused a loss of function of a previously existing gene, the regulator gene of the ribitol pathway. This mechanism has many similarities with the better-known lactose operon of *E. coli*, which allows the bacteria to use lactose as a source of sugar.

In summary, in both the human lactase gene and in the xylitol examples, one notices that there is no structural gene mutation. The protein or enzyme remains the same in the wild type and the "mutant." No new structures have been erected, but a change in a regulatory gene (promoter region) allows a preexisting enzyme to be produced under different circumstances.

The Sickle Cell Mutation

Another mutation considered to be beneficial to humans is the condition of sickle heterozygosity, of which the homozygous phenotype is sickle cell anemia (SCA). Sickle cell anemia is a recessive genetic disease caused by the replacement of the normal hemoglobin A with hemoglobin S (for "sickle"). Hemoglobin is an organic compound found in the red blood cells of many organisms and mainly serves to transport oxygen from the lungs to the tissues. Adult hemoglobin is a tetramer—that is, made of the assembly of four subunits, each made of a globin protein chain covalently linked to an iron-binding, non-protein cofactor called heme. The iron ion of the heme group carries the oxygen molecule. The protein tetramer is itself made of two alpha- and two beta-globin subunits.

The sickle mutation happens in the gene responsible for the synthesis of the β-globin protein subunits. A single nucleotide mutation at position six of the β-globin gene found on chromosome 11 results in the substitution at that position in the protein sequence of the correct amino acid (glutamic acid or Glu) by another one (Valine or Val). Homozygotes (SS) patients have two copies of the abnormal gene. They suffer high morbidity and mortality from the disease itself.[20, 21] Survivors lead an extremely poor quality of life, even worse in developing countries. In some situations, including those that lead to a low oxygen level, the red cells of affected patients lose their round and biconcave shape to adopt the shape of a sickle, and they become stiff and inflexible. The sickle shape is responsible for their premature destruction and the occlusion of capillaries, the smallest blood vessels of the body, causing respectively anemia and painful crisis. Any form of malaria, including *Plasmodium falciparum* malaria, worsens the toll on sickle cell patients. Carriers, on the other hand, have only one copy of the abnormal gene and are thus heterozygotes (AS). They do not have the disease and lead a normal life.

It has been found, however, that sickle cell carriers are less likely to develop the severe forms of falciparum malaria or to die from the disease than the genotypically normal patient (AA). They are less likely to develop the two most severe forms of falciparum malaria, namely cerebral malaria and malaria with severe anemia.[22, 23] Consequently, in populations where falciparum malaria is endemic, the prevalence of AS subjects increases at the expense of the AA genotype. A finding in congruence with this axiom of population genetics that there exists a close correlation between the prevalence of a particular variant within a population and its fitness, understood here as the ability to survive and produce offspring.[24] Case in point, the adult prevalence of the sickle cell trait throughout Nigeria is about one in four people, while the highest regional figure for the hemoglobin C trait, another similarly inherited hemoglobin anomaly but without the benefit of such a positive selection mechanism, is about one in seventeen.[25]

Sickle cell carriers demonstrate improved fitness compared to the normal genotype AA in an environment where falciparum malaria is endemic. However, this generational advantage is obtained at a greater cost for subsequent generations. In the hypothetical situation where the AA gene pool is not increased significantly enough through immigration or breeding with other populations, there will be a progressive increase in the prevalence of SS patients, and a decrease in the prevalence of genotypically normal (AA) subjects. The increased prevalence of the SS genotype is also made at the expense of the AS constitution. The sickle cell mutation is then maladaptive in view of the neo-Darwinian framework of evolution. The estimated prevalence of SCA in Nigeria, where falciparum malaria is endemic, is over 120 cases per 10,000 people, while the mean prevalence in European countries of cystic fibrosis, also a recessive disease but again without a positive selection mechanism of its associated heterozygous phenotype, is less than 1 case per 10,000 habitants.[22, 26] Of course, this is only if one excepts Ireland, an outlier in the comparative prevalence of cystic fibrosis, where intermarriage within the same gene pool is believed to have skewed the numbers.

All things being equal, patients with sickle cell disease fare much worse than those without when infected with malaria. The sickle mutation cannot be considered beneficial, because patients affected by the disease, including pregnant women, suffer great morbidity and mortality. A large meta-analysis of maternal and perinatal outcomes in pregnant women with sickle cell anemia compared to pregnant women without the disease shows that affected women were at a greater risk of maternal mortality and adverse perinatal outcomes such as miscarriages, stillbirth, preterm delivery, and the birth of infants said to be small for gestational age.[27-30] In short, the sickle mutation is all the way around detrimental, and the small benefit conferred to AS individuals is short-lived when viewed in both the long term and within a neo-Darwinian perspective, a theory it is supposed to buttress.

Adverse Outcomes of Hereditary Mutations

In most cases, random germline mutations are detrimental. The medical literature is replete with examples of germline DNA mutations that either cause or increase the risk of a number of hereditary disorders. One needs only to think, for instance, of the mutations that increase cancer incidences, such as some of the BRCA1/2 gene mutations, or the mutations that predispose to the hereditary colorectal cancer syndromes. Significant disability is also brought about by DNA mutations, causing noncancerous diseases such as sickle cell anemia, cystic fibrosis, neurofibromatosis, muscular dystrophy, etc.

The Origin of Eukaryotic Lineages

We have, then, made clear our rejection of random DNA mutations as the ferments of eukaryotic evolution. In the vast majority of cases, they are harmful to the individual organism and its descendants. Besides, the theory of natural selection does not provide the details of the clean splitting of one taxon into two distinct species. Some scientists have attempted, however, to fill this gap with the allopatric speciation theory formerly referred to as speciation by geographic isolation. This theory, initially propounded by Moritz Wagner,[31] and later developed by other biologists such as Ernst Mayr, has not been proven in the facts. For instance, human populations isolated for long periods by large oceans, or deserts like the Sahara, have not differentiated into different species or subspecies incapable of mating amongst themselves and producing fertile offspring, a counterargument also supported in our opinion by the extensive gene flow found amongst Darwin's subspecies of finches in the Galapagos archipelago.[32] A similar situation of hybridization is found in the "adaptive radiations" of cichlid fishes in the great lakes of Africa, such as the Malawi cichlids, and in the different subspecies of gulls in the Arctic Circle.[33-37]

We assert that the making of the subspecies of finches of the Galapagos Islands, the cichlids of the great African lakes, or the gulls of the Arctic Circle, bears similarities with the differential evolution of human phenotypes based on geography. A genome is in part the product of an ecosystem. A foreign or primary genome introduced into a new environment will be modulated by contributions from said ecosystem to form a second-generation genetic heritage without losing the master sequence characteristic of its initial identity. A phenotype in that new environment is the expression of that reformed genotype. The human races do not constitute distinct species; neither do the finches of the Galapagos, the gulls of the Arctic Circle, or the cichlids of the African lakes. The diversity within each category represents geographic phenotypes of the same species, and within each species, by definition, there are no reproductive barriers.

In other words, what Darwin had observed and reported from his studies of the Galápagos finches he had also witnessed among the various people he had encountered during his voyage aboard the *HMS Beagle*, some of whom he later described in his journal.[38] From the Englishmen he left behind at Devonport to the multiple specimens of the human species that fate had placed on his path during his five-year journey around the world, such as the Gauchos of las Minas, the face-painted natives of Tierra del Fuego half-clothed in guanaco hides, the tattooed, broad-shouldered Tahitians, the emu-dancing White Cockatoo men of Australia, and the tortured Black slaves of Pernambuco, the budding naturalist seems to have gone through a vibrant experience of intraspecific biodiversity, long before his famous meeting with the Galápagos finches.

Toward a New Model of Eukaryotic Biodiversity

The Seeds-First Theory

To explain eukaryotic biodiversity in nature, we will refer to a more encompassing theory we call the seeds-first theory. Without naming it, we have already drawn its first sketch in the previous pages. The seeds-first theory of eukaryotic biodiversity states that all the lineages of eukaryotes that have ever lived on Earth arose from distinct zygotes which, themselves, have resulted from the pairing of bacteria with different quasispecies-linked strains of viruses, in accordance with the bilateral descent theory of eukaryogenesis. The seeds-first theory does indeed treat oak shrubs, flamingo chicks, little lemur pups, and human babies respectively as members of genuine biological lineages that have stemmed from different original zygotes. Altogether, it moves the floral metaphor of eukaryotic biodiversity from the Darwinian branched tree to the vast cornfield of discrete lineages in parallel evolution.

Overview of the New Model

Here and in the next chapter, we will make the argument that the eukaryotes, once on the scene, have not diverged, whether according to a Darwinian or Haeckelian model. There is not a eukaryotic tree of life. We argue that the domain Eukaryotes appeared as multiple discrete lineages that evolved with time from one configuration to the next. Indeed, at any time during their history, eukaryotic species have always represented distinct entities revealed on a cross-section of the living world, while their real essence was in fact the present configuration of lineages that started as zygotes produced under the bilateral descent system. What we call a eukaryotic species is the temporal configuration of a certain lineage. It is a construct designed to capture the differences in behavior and appearance within each lineage. In other words, except at the hands of the hominin lineage, species or lineages do not go extinct; they evolve or change in appearance. We will arbitrarily name each lineage according to the current appellation of their configuration. We will, for instance, refer to the human, horse, apple tree, or mouse lineages. The human lineage would include Homo sapiens sapiens, Homo erectus, Australopithecus afarensis, and so on, including our aquatic configurations. Some skeletal remains that we have considered as fossils of different species may in fact only represent contemporaneous geographic variations of the same lineage.

The Darwinian concept "origin of species" comprises two different processes that must be disentangled. The first is eukaryogenesis, or the emergence of the eukaryotic lineages. For each lineage, the original configuration is also the original species. The second process is evolution, when, with time, a species changes into a new one. Speciation, as it has been understood in the Darwinian framework, then includes both eukaryogenesis and evolution.

Theory of Enzymatic Innovation

We have then been led to reconsider eukaryotic biodiversity under the light of our viro-bacterial theory of eukaryogenesis. The enzymatic theory of biodiversity, also referred to as the theory of enzymatic innovation, postulates that the quasispecies of RNA-based life, the families of DNA viruses, the prokaryotic and eukaryotic lineages all stemmed from the diversity-generating action of RNA-dependent polymerase enzymes when copying the RNA template. The reverse transcriptase enzyme, an RNA-dependent DNA polymerase, finalized the process of quasispeciation, leading to eukaryotes.

We take the position that the answer to the origin of eukaryotic biodiversity lies within the reproductive modality of RNA systems. The language of the genetic code established a relationship between an RNA codon and a translation product, which could be either an amino acid or a stop message. This language never directly involved DNA, which at any rate came later to the scene as an improved storage and safeguarding form of the heredity molecule, hence the many DNA repair mechanisms, including the exonuclease function of DNA-dependent polymerases. RNA-dependent polymerases, in contrast to DNA-dependent polymerases, lack a proofreading mechanism when copying the genetic material. However, what has been labeled as "error-prone" RNA replication not only is known to be beneficial to RNA viruses by forming clouds of viral quasispecies, but also fits with our theory that eukaryotes originate in part from specific RNA viruses of bacteria or retrophages. We contend that eukaryotic biodiversity originally stemmed from the lack of fidelity, or better, the innovative propensity of the RNA-dependent polymerase enzymes that replicate RNA genomes, principally the reverse transcriptase enzyme. The terminology used in current literature about the diversity-generating property of RNA-dependent polymerases is by and large negative, with terms such as "error-prone," "mistakes," "lack of fidelity," etc. We will continue to use those terms for the moment, knowing they are misnomers. For instance, Bebenek and colleagues have pointed out the high specificity and nonrandom distribution of the mutations associated with the HIV-1 reverse transcriptase enzyme.[39] Moreover, the interdependence of the lineages that resulted from the action of those enzymes disproves the notion that the mutations they also engineered were done by mistake and without purpose. For many reasons, we believe that biological processes are directed. The coding of genetic information at the origin of life to produce structures coevolving with the physical environment (Gaia theory*)[40], the degeneracy and the nonrandom codon allocation pattern of the genetic

---

* The Gaia hypothesis, initially conceived by the chemist and inventor James Lovelock (1972) but co-developed with biologist Lynn Margulis, postulates that biological systems and the physical environment are interdependent, as they

code,[41] and the parallel and coordinated evolution of flowering plants with their pollinators are some of the many examples that arguably betray "foresight" and purpose.

The RNA-dependent polymerases are enzymes responsible for the synthesis of genetic materials from an RNA template. They either facilitate the synthesis of RNA sequences from an RNA template or the conversion of RNA into DNA. They include the RNA-dependent RNA polymerases (RdRP) and the RNA-dependent DNA polymerases (RdDP), or reverse transcriptase. The mutation rates of viral RNA polymerases are in the order of 1 mutation per 10000 nucleotides copied.[42] Besides, the identification of 4 conserved motifs among the RNA-dependent polymerases strongly suggests a shared ancestry.[43]

As previously mentioned, RNA-dependent polymerases, unlike DNA-dependent polymerases, characteristically lack a proofreading mechanism and thus are of exceptionally low fidelity.[42, 44] This apparent handicap translates to a high mutation rate of RNA life forms. This is a well-known problem particularly when it comes to the difficulties associated with the management and control of infections associated with RNA viruses such as influenza, HIV, and Ebola, to name a few. This variability allows these viruses to rapidly adapt to vaccines and antiviral drugs. In contrast to the mostly deleterious effects of DNA mutations in multicellular organisms, those RNA viruses benefit from their genetic mutations, provided their rate is kept under a certain threshold. As a result of the low fidelity of the RNA-dependent polymerases, organisms with RNA genomes replicate with extremely high mutation rates, thus forming after several rounds of replications a collection of variants or a cloud of mutants called quasispecies.

---

function as a self-regulating negative feedback loop. This particular enunciation of the theory will be reviewed later.

RNA-Dependent Polymerase Enzyme Directed Speciation

Quasispecies theory, a population-based mathematical framework initially formulated by Manfred Eigen to explain the "self-organization and early evolution of life," is currently used to characterize the evolutionary dynamics of viruses with RNA genomes.[24, 45] In this context, Vignuzzi and colleagues explain that due to high mutation rates, an RNA virus population is not genotypically homogenous, but forms instead a cloud of related variants called quasispecies.[46] As also defined by Lauring and Andino, a quasispecies "is a cloud of diverse variants that are genetically linked through mutation, interact cooperatively on a functional level, and collectively contribute to the characteristics of the population."[24]

Within the limits of an error threshold above which a population can be led to extinction through "error catastrophe", genomic diversity is actually a strength of quasispecies, as it affords the population some plasticity, or a greater chance to evolve and adapt to the challenges of a dynamic environment. Cooperative interactions among the diverse variants of a viral pathogen can potentially lead them to a degree of virulence unreachable even by the most virulent variant clone in isolation. In fact, this was the result of a study led by Vignuzzi, who demonstrated that, when inoculated into naïve mice, a neurovirulent clone of poliovirus within a genetically constrained population could not enter the central nervous system (CNS) of the rodents and replicate. However, the brains of the injected mice were accessed, allowing for the replication of the neurovirulent clone only when it was inoculated, along with a diverse quasispecies of poliovirus.[46] Cooperation among the many variants of the diverse quasispecies most likely facilitated the entry of the neurovirulent clone into the CNS of the inoculated mice. Hence, in the context of viral quasispecies, population diversity is a determinant of strength and virulence. The quasispecies of RNA viruses strike a delicate balance between the error catastrophe of genomic mutation rates above a certain threshold and the rigidity of a system with high replicative fidelity, such as DNA. The benefit is a degree of variability and adaptability, constituting a survival mechanism that allows them to evade their host's innate defense mechanisms and overcome the effects of vaccines and antiviral drugs. In other words, this diversity-generating mechanism constitutes a kind of insurance policy for the survival of the viral master sequence in a dynamic environment, or when there is a shift in "the fitness landscape."

The diverse members of a viral quasispecies cooperate to achieve the goal of replication. Although they are different, their genetic information is organized around a stable master sequence that unites them. This strategy allows the cloud of mutants to adapt to environmental challenges while the master sequence is carried forward with time as part of a living entity.[24] The outcome of evolution is not adjudicated on the fate of an individual element of the quasispecies but on that of the entire population. A look at the reproductive biology of two RNA viruses, influenza A virus and HIV, may help us illustrate the formation of quasispecies.

Influenza A Viruses

Influenza A viruses (IAVs) are enveloped RNA viruses with an eight-segment genome. They are one of five genera of the family of viruses called Orthomyxoviridae. While the natural reservoirs of the IAVs seem to be wild aquatic birds,[47] those viruses infect a broad selection of hosts from humans and pigs to horses and birds.[48] They contain two important surface glycoproteins that play a major role in their pathogenesis. The hemagglutinin (HA) attachment protein binds to host cell receptors and mediates viral entry by fusion, while the neuraminidase (NA), on the other hand, facilitates the cellular release of newly replicated viruses and their cell-to-cell spread. Therefore, antibodies to the HA glycoprotein prevent receptor binding and thus neutralize the infectivity of the virus, while antibodies to the NA prevent the efficient release of the virus from infected cells.[49] IAVs display rapid evolutionary dynamics. Their year-to-year variations make them moving targets for preventive and therapeutic antiviral strategies.

One method of genetic diversification of the IAVs is antigenic shift, a type of reassortment of viral segments between two different strains of viruses. This method does not involve the polymerase enzyme. Coinfection of one host cell with two or more subtypes of IAVs can result in reassortant virions containing segments of more than one parent viruses. When the genetic reassortment involves the gene segments encoding one or both envelope glycoproteins, the process is called antigenic shift.[50] Reassortment between animal and human strains can create a new virus able to replicate in humans and cause pandemics with disastrous results.

But what concerns us here is the enzymatic diversification through RdRP. IAVs, like most RNA viruses, have very high mutation rates, "ranging from 1.8 to 8.4×10^{-3} substitutions per site per year (subs/site/year)."[48] Due to the low fidelity of viral RNA polymerases, IAVs exist as populations of quasispecies, and they inherently display high error rates.[50] Mutations in some viral strains that cause amino acid changes in the antigenic portions of their surface glycoproteins HA and NA may allow them to evade preexisting host immunity, which is a substantial advantage. Such mutations are called antigenic drift.

HIV Biology as a Model

Using the biology of HIV as a model, arguably our best-known retrovirus, we may further illuminate the concept of quasispecies.

Some in vitro studies have documented the lack of fidelity of the HIV-1 reverse transcriptase enzyme in the synthesis of DNA from the viral genomic RNA template. In one study in particular, the investigators found an average of one error per 1,700 nucleotides incorporated into a growing DNA chain.[51] Those errors can be of different types, including base substitutions, additions, and deletions. In some sequences called mutational hotspots, the error rate can top 1 per 70 polymerized nucleotides. The extraordinary diversity of the

HIV-1 genome is known to stem from error-prone reverse transcription. Another study reported single-base frameshift mutational hotspots and showed that RT-induced errors were non-randomly distributed.[39] The researchers further noted that, if effective in vivo, these errors would generate about 5 mutations per genome during each round of replication. The duration of a round of replication or generation time of HIV-1, defined as the time between a virion exiting a cell and it causing the release of a new generation of viruses through the infection of another cell, lasts 2.6 days (62.4 hours) according to a mathematical model in one study,[52] and roughly 2.17 days (52 hours) in another.[53]

This replicative modality implies that any two viruses sampled from an infected host are more likely to differ from each other. When the early HIV researchers first noticed the sequence diversity of viruses isolated in individual patients, they thought it was due to infections by different strains. However, Manfred Eigen then referred to the work of Simon Wain Hobson of the Institut Pasteur in Paris who demonstrated that the diverse HIV sequences were in fact related to each other. This work, according to Eigen, is a confirmation that RNA viruses, especially HIV, are quasispecies.[54] The RT-induced viral mutation and diversification represent a survival strategy of the virus itself. Likewise, the wide diversity of eukaryotic lineages is the transposition to a different level of this survival strategy. The continued blossoming of eukaryotic life on earth ecosystems despite the many extinctions of lineages caused by human activities and several other threats to life epitomizes this strategy at work.

The RT Enzyme, a Major Driver of Biodiversity Before Eukaryogenesis

RNA-based life has then diversified tremendously under the guidance of the RNA-dependent polymerases with every single replication, from the earliest life forms through the RNA cells, and the retrophages attached to the receptors of their prokaryotic hosts. That process continued with the reverse transcriptase enzymes up until the definitive incorporation of the prophage DNA into the cellular genome. These quasispecies or clouds of variants of RNA systems form the basis of current subviral, viral, prokaryotic, and eukaryotic biodiversity. It is then through the properties of the RNA-dependent polymerases, namely their innovative propensity or so-called "lack of fidelity," that there has emerged a cloud of RNA-based organisms, which will lead to spectra of retrophages and their associated bacteria.

One strain of retrophages, however, will be the common viral ancestor of the animal lineages through the formation of quasispecies by repeated rounds of replication. This is formed when a subset of retrophages from the population resulting from each round of replication reinfect the *Thiomargarita namibiensis* bacteria until the last branch of the quasispecies. This process will create, through the reverse transcriptase enzyme, a cloud of related variants ready now to fuse indefinitely with the bacterial hosts. The bacteriophage DNA will subsequently transform its metabolically sluggish gammaproteobacterial host into a eukaryotic cell. It is the common proceeding for the emergence of the four eukaryotic kingdoms, namely Animalia, Plantae, Fungi and Protista. For instance, similar processes happened with the bacteria of the respective genera *Beggiatoa* and *Thioploca*, forming the lineages of the flora. The phylogenetic "Tree of Life" is made up of organisms with RNA genomes.

The reverse transcriptase enzyme set up eukaryotic phylogenesis in each kingdom as it transcribed the retrophage RNA information into DNA. The retrophage genome, we recall, will contribute the viral half of the genome of the first eukaryotic cells. Eukaryotic biodiversity is, at its root, the result of enzymatic quasispeciation through "error-prone" replication of RNA genomes. The diversity of life forms such as revealed in the fossils of the Late Precambrian and Cambrian geological periods was the eukaryotic reflection of clouds of retroviral quasispecies in symbiosis with their bacterial hosts.

The genomic differences across a snapshot of eukaryotic lineages stem originally from the action of the diversity-generating RNA-dependent polymerases, leading to a diverse array of retrophages and bacteria. The RT enzyme will again diversify the retrophages during reverse transcription. Further down the process, metazoans will generate the first eukaryotic gametes through meiosis. These gametes will bear the genomic and phenotypical differentiation features of their progenitors and will continue to coevolve within each lineage. In fact, considering the wide interspecific morphological variation of mammalian spermatozoa, some have used sperm morphology as a phylogenetic trait within that class.[55]

The apparition of various complex life forms during the Late Precambrian and Cambrian periods represents the transposition into macroscopic and fossilizable eukaryotic organisms of genetic scripts and their catalysts that have evolved since the beginning of life in submicroscopic and non-fossilized living entities, the Cambrian animals having evolved, in a saltatory manner, from the Precambrian fauna. This development was initially made possible by the merging of information contained within viral and bacterial protagonists. In the mass of the multiform organisms of the Cambrian, all the lineages were represented and have since remained whole. Even us, humans, our own ancestors were also there in the Cambrian. The ancestors of eukaryotes on one side of the bilateral descent system were viruses already differentiated into quasispecies. The truth is that during the Late Precambrian period, Earth became pregnant with new forms of life. The different geological layers represent chronological radiographs of its hermaphrodite belly. The periods that appear to have been lifeless to the paleontologist are the films taken during the successive processes of gametogenesis, fecundation, and early embryogenesis. The growing boneless embryos at first, and later, the fetal bones, even when their respective projections left the impression of suddenness, are nonetheless anticipated and historical.

Indeed, the earliest eukaryotic fossils of the stratigraphic column belonged to organisms conceived from the merging of minuscule and hardly fossilizable ancestors namely viruses and bacteria that have evolved within Earth itself. Likewise, the abdominal scan B of, say a 5-month pregnant woman, will show fetal bones indicating the presence of an intrauterine vertebrate organism, an unforeseeable fact for a physician if his only information is a scan A taken of the same woman 5 months earlier. For illustration's sake, let's say that Scan A was taken a few hours after the woman had both ovulated and engaged in the sexual encounter responsible for the pregnancy, which means that the sperm and the egg from which the fetus originated were already in the woman's genital tract. The gamete cells would not have been seen on Scan A because not only are they not as radiodense as the bones of the fetus, but they are also microscopic in size.

However, scientists may have identified in the sedimentary rock strata some of the embryos that have resulted from the fusion of viruses and bacteria. For instance, consistent with our seeds-first theory, we would not be surprised if Chen and co-workers were correct when they reported in April 2000 what they identified as " putative" fossil sponge embryos and larvae from the Precambrian Doushantuo Formation in Southwest China.[56] And if similar microfossils are found in Late Precambrian rocks around the world, neither should we. That is, the seeds-first theory predicts such findings. Likewise, we claim that most, if not all, of the embryos of the animal lineages were formed during this geological period and that they displayed a configuration similar to that of a sponge embryo.

Lastly, given the ecology of bacteriophages and of sulfur bacteria, this analysis lends credence to the contention that hydrothermal systems have played an essential role in the origin of life. Quasispecies of eukaryogenic retrophages foretold the subsequent countless number of eukaryotic lineages. A lineage is complete with the final transcription into DNA of the viral RNA and its hybridization with the bacterial host genome. Further genomic modifications during evolution will not deviate from this general outline. Eukaryogenesis resulted, according to the bilateral descent theory, from the rendezvous of metabolically slow bacteria and the reverse-transcribed genomes of a collection of genetically linked temperate retrophages. It sealed once and for all the fate of some retroviral quasispecies as progenitors of a new domain of life. If we can summarize our pedigree as virus-bacteria hybrids, we recognize that phages have played a prominent role in eukaryotic biodiversity.

On the Ubiquity of the Eukaryotic Lineages

The biogeography of the bacterium *Thiomargarita namibiensis* seems also to give a clue about how the lineages managed to colonize different regions of the world. This bacterium, besides its initial discovery in Namibian shelf sediments, has also been found, along with other morphotypes, at several places around the world. Those places are as diverse as the Gulf of Mexico; the Pacific margin of Costa Rica; the Neoproterozoic Doushantuo Formation in South China; Hydrate Ridge, offshore of Oregon; and mud volcanoes of the Barents and eastern Mediterranean seas.[57, 58] It is then obvious that the oceans have disseminated around the world samples of the bacteria at the origin of the animal oocyte; but also in the context of plate tectonics dynamics forming and breaking off supercontinents, the spread of the lineages all over our planet is quite understandable. This realization undoubtedly provides an alternative to the proposition that migration by land and makeshift rafts explain the presence of most lineages throughout Earth. Humankind, for instance, did not make the proverbial Paleolithic journey from Africa to colonize the far corners of the world.

We must remember that all life forms have colonized the lands from the oceans, and perhaps also from bodies of fresh water. How our own aquatic ancestors made the transition to bipedal tellurians can be guessed but is not known. From virus-bacteria hybrids, they ultimately evolved to their last aquatic configurations. They would then be washed up onto the shores of landmasses, from large continents to small atoll islands. Different configurations of the human lineage, for instance, could have evolved in those different places without much more migration than that which comes with such basic phenomena as wars, hunting and gathering, a pastoral lifestyle, natural calamities, etc. The biogeography of *T. namibiensis* and of *Tn*-like bacteria do not seem to disprove such a theory. The bipedal ancestors of European or Asian populations unlikely journeyed from Africa, as has been described in the scientific literature. We surmise that they have evolved *in situ* from aquatic ancestors. It might just be that the oocyte that is our common maternal heritage, the bacterium *Thiomargarita namibiensis,* first evolved off the coasts of what is Africa today. The OvaHimba tribes of West Africa may be among those keeping this "memory" alive in a very subtle way.

For instance, the OvaHimba homestead is an oval-shaped living arrangement, usually for an extended family, centered by a livestock enclosure or kraal and bordered by human dwellings and a fence (Fig. 7). An ancestral or holy fire through which the villagers communicate with their ancestors, and with God — whom they call Mukuru — is located in the middle of a line that goes from the main hut to the kraal. This living arrangement is reminiscent of the anatomy of an egg, with the kraal representing the yolk and the holy fire the germinal disk that holds the genome. The central fireplace is hallowed ground; it connects the villagers to their ancestors as does the genome for a cell.

Figure 7: An OvaHimba homestead. A kraal is seen at the center of the village.
Attribution: https://www.youtube.com/watch?v=blBkC4j6hYk
Courtesy Cody Buffinton.

CHAPTER

5

Mechanism of Eukaryotic Evolution

> I stood flapping my arms, first the left and then the right, as I had seen Pamphilë do, but no little feathers appeared on them and they showed no sign of turning into wings. All that happened was that the hair on them grew coarser and coarser and the skin toughened into hide. Next, my fingers bunched together into a hard lump so that my hands became hooves, the same change came over my feet and I felt a long tail sprouting from the base of my spine. Then my face swelled, my mouth widened, my nostrils dilated, my lips hung flabbily down, and my ears shot up long and hairy . . . At last, hopelessly surveying myself all over, I was obliged to face the mortifying fact that I had been transformed not into a bird, but into a plain jackass.
>
> —*The Golden Ass*. Robert Graves' translation of Lucius Apuleius's *Metamorphoses*.

Much has been made of the Cambrian explosion, the relatively sudden apparition in the Cambrian geological layer of major animal groups. However, impressive as it was in number and complexity, such as initially revealed in the legendary Burgess Shale and later in the Chinese Maotianshan, the Cambrian fauna had direct ancestors in the deeper geological layers. To borrow an analogy from the world of technology, the Cambrian explosion was a major delivery to market of fossilizable products whose genetic programs and hardware had taken the universe billions of years to devise and evolve.

While the transitional process from prokaryotes to eukaryotes had remained largely obscure until now, significant efforts have been made, particularly in paleontology, to try to pinpoint its timing. Microstructures interpreted as fossils of cyanobacteria and identified in a 3.5-billion-year-old sedimentary rock in Western Australia (Warrawoona) have been considered, albeit with much controversy, the earliest record of cellular life on earth.[1] Among the oldest evidence of eukaryotic life are 1.5–1.8-billion-year-old microfossils that have been discovered in shales in Northern Australia and Northern China.[2-4] Javaux and colleagues have reported having extracted fossils of putative eukaryotic microorganisms from rocks more than 3 billion years old.[5] The biological affinity of those organic-walled eukaryotic microfossils cannot usually be determined with certainty, and when they are 50 micrometers or more in size, they are called acritarchs.[6, 7]

However, it was not until the end of the Neoproterozoic era, about 600 million years ago, that multicellular organisms or metazoans appeared in the geological record, as revealed by the Ediacaran fossils discovered in almost all the continents, but precisely in places as diverse as Australia, Namibia, China, England, Siberia, Newfoundland, etc.[8] Just to take a few examples, the animals of the Ediacaran period such as Dickinsonia, Charnia, Spriggina, and Parvancorina display no direct similarities with the trilobites, brachiopods, or Hallucigenia of the Cambrian. The many dinosaur species of the Triassic period 240 million years ago (i.e., the Mesozoic era) are not represented as such among current vertebrates or those of the preceding Paleozoic era. It ensues that evolution (at least in the sense of phenotype changes through time) is an undeniable fact of life.

Biological evolution does also present some similarities with the physical evolution of matter from quarks to the nucleus, from hydrogen atoms to stars, with the intermediate stages of clouds of matter in "stellar nurseries" and protostars.[9, 10] The controversy about biological evolution is mainly about its meaning, mechanisms, and driving force. For instance, the proposed mechanism went from the inheritance of acquired characteristics with Lamarck to natural selection acting on "descent with modification" with Darwin, or on genetic DNA mutations with the neo-Darwinists, and so on. Until now, the major determinant of eukaryotic biodiversity was unfortunately confused with phyletic evolution.

We have demonstrated in the previous chapter that biological diversity was initiated through an enzymatic mechanism. Biodiversity is rooted in the nonrandom modification of nucleotides during the replication of RNA systems, as spearheaded by RNA-dependent polymerase enzymes, including the reverse transcriptase. Phylogenesis explains, for instance, the differential origin of a man (*Homo sapiens*) and a dog (*Canis lupus familiaris*). The concept of evolution on the contrary, again, has been squandered and used with different acceptations. Therefore, much confusion about it has been created. Building on Dobzhansky's definition,[*11] we will say that evolution is a change, with time, in the genetic and epigenetic information of a group of conspecific organisms, resulting in a difference in phenotype from time A to time B. Evolution does not use one single mechanism for all life domains (eukaryotes, prokaryotes, RNA systems, DNA viruses). Each major domain of life evolves according to its own, though related, modality. The evolution of RNA systems is the model of speciation par excellence and unfolds in a tree-like pattern, hence the quasispecies concept. It is mediated by RNA-dependent catalysts, from the earliest RNA polymerases

---

* "Evolution is a change in the genetic composition of populations."

to the reverse transcriptase enzyme. DNA systems, whether viral, prokaryotic, or eukaryotic, evolve by their own means.

Briefly, the process of formation of the eukaryotic lineages is completely separate from their evolution. The former obviously preceded the latter. Evolution here does not mean the splitting of a taxonomic group into two or more different subgroups, as in a Darwinian or Haeckelian evolutionary tree. There are no forks on the evolutionary trajectory of eukaryotes. Evolution means the process it took to transform over time the genome and the phenotype of a eukaryotic lineage from its emergence to its modern characteristics. During a specific snapshot of time, it can mean the evolution, for instance, of *Australopithecus afarensis* to *Homo sapiens*.

The natural evolutionary boundaries of a eukaryotic lineage are essentially dictated by the characteristics of the genome of its last viral ancestor, as modified by the reverse transcriptase enzyme, right before the final merging with its corresponding bacterial host genome. Enzymatic viral quasispeciation is then understood to have laid the blueprint for the direction of evolution for each eukaryotic lineage. Once formed, the lineages will evolve in a parallel fashion, regardless of the time of their apparition. It follows that the evolutionary vectors of at least the interdependent lineages are equal, that is, of the same direction and magnitude. For instance, pollinator-dependent plants coevolved with their pollen vectors; the same goes as well for some other plants and the frugivorous birds they depend on to disperse their seeds. There are many examples of these mutualistic relationships in nature.

Moreover, the deep-throated, trumpet-shaped red flower of the plant *Brugmansia sanguinea* did not predate the long and sword-shaped bill of the hummingbird *Ensifera ensifera*; they coevolved. In other words, this variant of hummingbird did not evolve a long beak to "adapt" to a preexisting red angel's trumpet or the other way around. The two protagonists evolved in a parallel fashion according to equal vectors, which ensured this continued mutualistic symbiosis over evolutionary time. The natural evolution of interdependent lineages is a running wide-belt treadmill on which the protagonists are positioned side by side or on parallel lines. No one lineage in an evolutionary duet unilaterally changes position relative to the other. It is as in Lewis Carroll's *Through the Looking-Glass* novel: "'Now, here, you see,' says the Red Queen to Alice, 'it takes all the running you can do, to keep in the same place. If you want to get somewhere else, you must run at least twice as fast as that!'"[12]

The Linear Mode of Eukaryotic Evolution

The evolution of eukaryotes is a linear phenomenon that takes a lineage from one phenotype at a defined time to a different one in the future. It can also be referred to as orthogenetic or straight-line evolution. There is no branching or splitting from a common eukaryotic ancestor to form two or more new species or lineages. A well-known example of this mode of evolution is that of the horse (*Equus caballus*) from its development as the dog-sized, three-toed hindfoot *Eohippus* of the Eocene to the modern one-toed *Equidae* equipped with a hoof.[13, 14] *Eohippus* and the modern horse are connected on a continuous evolutionary line by intermediary phenotypes such as *Mesohippus*, *Merychippus*, and *Pliohippus*, which in the framework of Darwinian natural selection are considered extinct species, but in reality, as we said, were only transitional configurations of one particular lineage, the horse lineage.

However, one does not need to look far for examples of orthogenetic evolution. The evolution of mankind, say, from *Ardipithecus* or *Australopithecus* to *Homo sapiens* in the African savanna is a case in point. It is a constant progress toward an increased sophistication of forms, skills, and behaviors. Contemporaneous hominins, while they may have shown some differences in skeletal anatomy, may not have represented different stages of evolution but rather different races or ethnicities. We refer here for example to the craniofacial anthropometric differences between the fossils of *Homo neanderthalensis* and *Homo sapiens* who lived contemporaneously for some time in Europe. Some of the craniofacial features of the Neanderthals included, among others, prominent brow ridges, a low and slanted forehead, a non-protruding chin, and an occipital bun.[15] Nonetheless, those archaic humans were not of a different lineage or species than their *sapiens* neighbors. There is a long history of classification of the human races or ethnic groups based on strict craniofacial anatomic differences.[16] Those differences, however, do not make them different lineages or species. In fact, the existence of statistically significant differences in facial anthropometric dimensions according to race and even sex (sexual dimorphism) has been confirmed by at least one study.[17] Jerry Coyne rightly observes in a different context that while indigenous populations look different from one continent to another, we do not label them as different species.[18] In his words: "The Inuit of Canada look different from the !Kung tribespeople of South Africa, and both look different from Finns. Do we classify all of these populations as different species?"[18]

Some scientists consider the Neanderthals, accurately in our opinion, as indigenous *Homo sapiens* who evolved during the glacial periods and thus were built specially to live in cold climates. Those who share this view refer to them as *Homo sapiens neanderthalensis* as opposed to modern humans (*Homo sapiens sapiens*). Their consideration to be of the same lineage as the hominins contemporary to them is also supported by the discovery that a certain percentage of the genetic makeup of modern humans is inherited from the Neanderthals. It was initially thought that only Asians and people of European ancestry inherited Neanderthal genes, but researchers from Princeton University, using data from the 1000 Genomes Project, have recently found that remnants of Neanderthal genomes have survived in every modern human population that has heretofore been studied, including people of African ancestry.[19]

The Tempo of Evolution

While the evolution of eukaryotes is a linear phenomenon, the role of the stratigraphic record in determining whether this evolution was uniformly slow and gradual, or occasionally punctuated by saltations or rapid morphological changes, cannot be overstated. The former modality termed "phyletic gradualism" by Eldredge and Gould[20] was initially articulated by Darwin, a backer of the Linnean maxim *Natura non facit saltum* (Latin for "nature does not make leaps"). Evolution would happen over long periods of time by small, cumulative, incremental changes resulting from random genetic changes acted upon by natural selection. In the latter modality, rapid macroevolutionary changes would happen episodically between long periods of stasis or evolutionary equilibrium, particularly in peripherally isolated populations. This is in part the essence of the theory of punctuated equilibria formulated by Eldredge and Gould during the 1970s to replace the exclusively gradualistic Darwinian approach.[20, 21]

Unfortunately, this theory, kindly nicknamed "punk eek," was also conceived within the natural selection framework and the invalid allopatric model of speciation. The gradualism vs. "punk eek" tug of war, however, is best resolved by discoveries coming from fields as different as paleontology and evolutionary genetics. The outcome is also likely to not be a winner-take-all situation. As an exhibit of a gradualistic evolutionary pace, Gordon Rattray Taylor offers the example of many lineages of bivalves that have not changed much in 400 million years of evolution.[13] On the other hand, we think the evidence exists for a theory of evolution occurring predominantly by alternating stasis and periods of rapid changes. As we will see later, the "punctuations" in the evolution of eukaryotes are caused, in all likelihood, by the coordinated expression of new sets of genetic and epigenetic information. It is the resulting novel organismic configurations that the paleontologist labels as new species.

Indeed, as Gould and Eldredge pointed out, "stasis is data," and the discontinuities in size and shape found in the fossil record may be telling us something.[21] The failure of the stratigraphic evidence to display the set of data expected by the gradualistic school of thought should not always be ascribed to the overused argument of its imperfections or its "incompleteness." The evolution from bacteria to eukaryotes, from fish to amphibians (tetrapods), from dinosaurs to birds, and most radically, the apparition in the Cambrian of new biological forms, may contain instances of saltations. In that sense, a more accurate picture of evolutionary tempo would be that the evolution of the lineages creates time-dependent morphological spectra.

On Faunal Extinction

With either of the possible rhythms of evolution, when the phenotype of a species has changed enough to be perceptible in the fossils they have left, the latter will leave us the wrong impression of extinction. We define an extinct species or lineage as one that does not leave any descendants. Faunal extinction, we believe, is essentially a hominin phenomenon. We are the only lineage that kills indiscriminately, willingly or not. Viruses and bacteria, in the course of their reproduction, may kill some of their hosts but never annihilate a lineage. They are often less lethal for certain demographics within a population. Prey animals often have a predilection for the weak, lame, and defenseless, such as the chicks or young calves in a group. Non-hominin animals generally kill for obvious biological reasons such as food, territory, family protection, or to increase the prospect of transmitting their genes to a new generation, etc. We of the hominin lineage do not spare. We kill for a variety of reasons, including for fun or for the unproven therapeutic effects of special cuts of certain animals, for instance. Moreover, we have pushed other lineages to the edge by the loss or contraction of their habitat that has resulted, for example, from our demographic expansion. Several researchers and analysts think that the hands of some of our ancestors were steep in the extinction of the woolly mammoth, for instance, most probably through an overkill mechanism.[22, 23]

The death of a given organism is intrinsic to the phenomenon of life and occurs naturally sometimes after the said organism has passed reproductive age. It is fundamentally different than the supposed extinction of a lineage. The biological theory of species extinction, a corollary of the natural selection framework, has no equivalent in physics or chemistry. Natural selection and the concept of species extinction introduce the prerequisite of trials and errors of nature, hence of waste. They propose that nature would evolve a whole lineage over several millions of years, and then in a geological click, for no apparent reason, would selectively dispose of its members, however geographically distributed, leaving only their fossils in the record. Apart from the traditional prevalence of this assumption in biological thought, there is nothing analogous in the rest of science. Nature has always been regarded as an excellent steward of its resources.* In physics, this idea is codified in the first law of thermodynamics, also known as the law of conservation of energy. Lavoisier's law of conservation of mass, according to which, in a chemical reaction, mass remains constant from reactants to products, renders the same principle in chemistry. One of the things that captivated my attention during a visit to the American Museum of Natural History in New York City in the course of my research for this book was an exhibit

---

* Reproduction, where, for an organism, the ratio of gametes that can potentially transmit their genetic material to the number of gametes produced is very low, may only have the appearance of waste.

in the dinosaur section, affirming the extreme rarity of bird fossils during the "Age of the Dinosaurs."

It instantly came to my mind that the emergence of birds must have had something to do with the so-called extinction of the dinosaurs. That would be saltation or punctuation after a long equilibrium episode in the evolution of the dinosaurs (although we know that this transition happened more smoothly than implied here). Moreover, we should point out that there is a growing body of evidence that most dinosaurs had feathers, and that many had hollow or pneumatic bones among other similarities with birds.[24-25] However, the German paleontologist Otto Schindewolf (1896-1971) is reported to have entertained the view that the first bird might have flown right off the nest of a dinosaur.[26] And, for as dramatic a depiction this is, it does reflect the true reality of the characteristically linear mode of eukaryotic evolution. It also illustrates the geographic isolation of a generation that took to the air while leaving their parents on the ground or at a lower altitude. For each lineage of dinosaurs that have evolved into birds, the last generation of ground-dwelling reptiles will be the last of that phenotype, and their carcasses will be known to us as fossils of an "extinct species" in the Darwinian model, but, in fact, the genes of the lineage would have been smoothly transmitted to a new configuration. Major saltations could have happened with the transition of a lineage from a type of environment to a different one, such as from sea to land or from land to air. We know some of those changes even in modern life, such as the transition of a chick from an anaerobic metabolism within the egg to an aerobic mode after hatching. The respiratory and cardiovascular changes of a newborn mammal are some other examples of macro-modifications with environmental changes. For instance, with the first cry of the human baby at birth, the process of gas exchanges is transferred from the placenta to the pulmonary system. Moreover, the right-to-left cardiac shunt via the ductus arteriosus that existed before birth is switched to a left-to-right shunt as a result of the now higher aortic pressure.[27] And we have not even explored metamorphosis as a possible modality of evolutionary saltation. The transformation of a larva (for example, a caterpillar) into a butterfly may be a mimic of a potential modality of evolution.[28, 29]

We then propose that, regarding the species which seem to have disappeared from a deeper geological stratum to that just above, we seek their descendants in the species of the more superficial stratum which were not present in that from below. In all fairness, the apparent relatedness between "extinct" and extant species was alluded to by Darwin in his book *The Voyage of the Beagle*: "This wonderful relationship in the same continent between the dead and the living, will, I do not doubt, hereafter throw more light on the appearance of organic beings on our earth, and their disappearance from it, than any other class of facts."[30] The observation, however, while on point, was interpreted differently than our proposition. It was presented as the result of a struggle for existence, or a competition between related species (or "species closely allied in habits"[31]) for the same limited resources, leaving winners and losers, respectively extant and extinct species.

The possibility of a real extinction can be seriously considered when the disappearance of a species from one stratum to the next cannot be explained by a dearth of fossils for one reason or another, or by complete migration to a new ecological niche. What have been called extinct species, for the most part, are successive or transitional skeletal phenotypes left behind in the fossil record by particular lineages as they evolved from one form to another. The magnitude of saltation is measured in the difference from phenotype A to phenotype B. It is up to scientists who spend a lifetime studying fossils to determine the pedigree of a living species from the database of heretofore considered extinct species, or what constitutes the descendants of the latter.

This leads us to ask the following question: What is then the mechanistic underpinning of evolutionary change within a eukaryotic lineage?

Evolution by Transposable Element (TE)-Induced Genetic Recombination

Each eukaryotic lineage has emerged individually from the fusion of a virus strain with a bacterium. Over time, the eukaryotes evolved in a linear and parallel fashion from their initial configuration to their place in the living world of today, older configurations buried in the geological layers giving the false impression of extinction. The different lineages of the animal kingdom have evolved from their quite similar sponge designs to species as externally distinct, for instance, as man, shark, and polar bear. It is as if animal evolution followed a pattern resembling that of ontogenesis, where the external appearances of the lineages evolved to the most unique from the most general. This is also quite evocative of the "biogenetic law" or the recapitulation theory that was inspired by the phrase "Ontogeny recapitulates phylogeny" by the German naturalist Ernst Haeckel. The ontogeny of a mammal, for instance, which starts with the fertilization of an egg, is followed by the nidation of the embryo and its subsequent development until birth. This process is mirrored in the formation of a virus-bacterium hybrid and the successive configurations of sessile or sponge-like animals (anchored on surfaces at the bottom of the ocean), fish, amphibian, proto-mammal,[32, 33] and mammal, characteristic of the evolution of this particular taxon. However, the question of evolution could be worded as follows: How did we get, for instance, to the modern man, the cat, or the apple tree from virus-bacterium hybrids?

Before we dive into the mechanism of eukaryotic evolution, we will first establish how we distinguish evolution from adaptation, two different concepts that are often used interchangeably in biological literature.

Adaptation

From the outset, we argue that the concept of adaptation is inadequate to characterize the overall interactions of living systems with the global terrestrial environment. Life does not "adapt" to Earth's environmental conditions; they coevolve. Living systems on Earth are not foreign phenomena juxtaposed to a planet whose conditions they need to adapt to. Life is a creation of the ecological conditions of our planet, and both subsystems are constantly engaging in different types of feedback interactions. A few examples will help to illustrate this point. Volcanic eruptions of the early earth had created a carbon dioxide-rich atmosphere by 2.7 billion years ago during the Neoarchean,[34, 35] which fostered the apparition of photosynthetic organisms.[36] Those organisms then produced the oxygen concentration necessary for the emergence of animal life and contributed to the formation of the ozone layer designed to protect life itself against the deleterious effects of solar ultraviolet radiation. In a more contemporaneous example, the international community now understands the impact of human activity on the global climate and is conscious of the need to take steps to reduce greenhouse gas emissions (for example, through the United Nations Framework Convention on Climate Change of 1992, the Kyoto protocol of 1997, and the Paris Climate Accords of 2015).

On a smaller scale, the negative feedback loop of algal blooms seems to support the Gaia concept. Continent-sized blooms of the photosynthetic phytoplankton *Emiliania huxleyi* are destroyed by the lytic Coccolithoviruses, forming massive milky oceanic shapes that can be seen from space. The visible bloom results from the release in the surrounding sea of the chalky shell of the algal cells. In the upper atmosphere, oxidation of the dimethyl sulfide (DMS) gas released by the dead organisms induces cloud condensation, which then blocks the sunlight, thus interfering with further phytoplankton growth.[37]

Life, while closely interacting with its environment, is also allowed within a relatively wide variation of climatic conditions. It is only that some phenotypes may experience a relative decrease in fitness or incur an additional fitness cost in non-original environments. For example, Black people living in temperate regions have on average lower vitamin D levels than their non-Black counterparts, with all the associated health disparity implications.[38, 39] Conversely, white people in South Africa, for instance, display one of the highest annual rates of malignant melanoma in the world.[40, 41] However, despite these setbacks, both human phenotypes can thrive in these "foreign" environments.

Therefore, adaptation of life forms at least to geographical variations of the global ecosystem does happen. We consider adaptive the reversible changes of an organism when there is also a change, often sudden, in the characteristics of its environment, such as occurs when an entity is moved from its natural habitat to a different one. We obviously reject the view that lineages evolve simply to adapt to changes in the environment. There are local differences in the evolution of species that may, for instance, account for the different human subpopulations or different beak shapes of finches, but environmental conditions do not engineer evolution. If that were the case, we might expect for ourselves a Neanderthal or Australopithecine regression in the future, should the ecological conditions under which these ancestors emerged return. We have erroneously considered the bones of our ancestors as relics of beings lacking the capacity to adapt to environmental changes.

Adaptive changes are well characterized and better understood in viruses and bacteria. They have allowed those organisms to adopt different behaviors in congruence with environmental contingencies. Some organisms adapt to environmental changes or respond to a new type of need by modulating the expression of certain genes. At any rate, the salient feature that permeates adaptation in the life forms where it has been studied is the central role of transposable or mobile elements.[42]

This is the case of the *E. coli* bacteria, for example, which characteristically do not produce the enzymes necessary for the transport and hydrolysis of lactose in an environment deprived of this nutrient, but do so when lactose is introduced. *E. coli* possesses a lactose (*lac*) operon of which the *lac* repressor protein is activated in an environment deprived of lactose, repressing the synthesis of the *lac* enzymes. Conversely, the repressor is inactivated when lactose is introduced. A transposon originally isolated from *Yersinia enterocolitica* carries a *lac* system that displays some homology with the *E. coli lac* operon.[43, 44] It is then possible that the *E. coli lac* operon may have originally been a transposon that later stabilized by losing its transposability. This is a quite similar situation to the mechanism of the previously reviewed lactase persistence in certain human populations.

As an example in the world of viruses, the cl protein operon system of the lambda bacteriophage operates in a fashion akin to the *lac* operon. Bacteriophage lambda does indeed have an operon-type system controlling its two functional states: lysis and lysogeny. The "decision" of the lambda phage to go lytic or to lysogenize after infecting its *E. coli* host depends on the concentration of a protein, the cl repressor, within the host itself. The metabolic or growth rate of the bacterial host inversely correlates with the concentration of the cl repressor protein. A low concentration of this protein induces the lytic program while a high concentration allows the bacteriophage to lysogenize, i.e., to lay dormant until better conditions arise. The lytic program is induced in more than 90% of cells with logarithmic growth, and thus with a low cl protein concentration. On the other hand, in metabolically inactive, stationary phase cells, there is enough cl to induce lysogenization in more than 90% of infected cells.[45]

Furthermore, a recent study illustrates a case of variation in gene expression as a mechanism of eukaryotic adaptation.[46] It identifies the involvement of transposable elements in regulating the synthesis by a fungus of melanin, a metabolite considered an adaptive trait for host colonization and survival in stressful environments. Transposable elements afforded to those fungi a degree of genomic plasticity that allows variability in melanin accumulation in function of need, as the production of this metabolite carries an associated fitness cost. The investigators demonstrated that variability in melanin accumulation was determined by the modulation of gene expression rather than through exonic mutations.

The argument of the fundamental role of TEs in the variability of gene expression is further illustrated by the finding that a transposable element is responsible for the establishment of polymorphism in the peppered moth (*Biston betularia*).[47] The story of the replacement in Britain during the Industrial Revolution of the common pale (*typica*) phenotype of the insect by a previously rare black (*carbonaria*) form has been hailed as a classic example of industrial melanism, and has represented an iconic illustration of evolution by natural selection.[48, 49]

However, the dark-colored form was engineered not by a point DNA mutation but by a transposition mechanism, and it remained the same species as the *typica* phenotype. It also appears that the effect of the *carbonaria* TE on melanization is achieved by altering the expression of a gene, the gene *cortex,* or more specifically by increasing the abundance of its mRNA transcript.[47] It remains that to our knowledge, it has not been established in the facts whether the reported increased prevalence of the melanic variety of the peppered moth brought about by industrialization was the result of its positive selection fostered by coal pollution, or a form of adaptation of the previously light-colored type under the pressure of increased predation.

Evolution of Eukaryotes

As we have demonstrated, the evolution of eukaryotes has not created a tree of species from a common eukaryotic ancestor as described in the Darwinian model. The family tree from which all life forms originate is instead based on RNA. Biological evolution, the impetus of which will be discussed later, is a vectorial process that takes into account the interplay between an organism's genetic heritage and information obtained from its environment. Evolution is linear, unidirectional, and irreversible. Lineages evolve in part according to information from the environment, large or small. Members of the same lineage that live separately in different environments over a long enough time will display, at any historical snapshot, intraspecific evolutive differences. Nonetheless, they still belong to the same species and thus remain interfertile.

According to our model, eukaryotic living systems form multiple lineages that have remained whole across time since their inception, rather than diversify according to a Darwinian or Haeckelian model. The human lineage spans several major phenotypic configurations, say, from our days as sponges, fastened to the bottom of the oceans until our evolution to terrestrial bipeds. The journey is similar but on a parallel line for any other mammals or birds, with the obvious difference that birds have different lines of ancestors. There is then a chimpanzee lineage, a gorilla lineage, a finch lineage, an *Arabidopsis thaliana* lineage, an apple tree lineage, and so on and so forth. The human lineage further evolved, say, from *Australopithecus* to modern *Homo sapiens*, through successive hominin configurations such as *Homo habilis*, *Homo erectus*, and different archaic *Homo sapiens* who likely were products of different environments. One species or configuration changed in time into the next one. Again, the large majority of specimens from the fossil record that are not represented *idem* in today's biodiversity are not from extinct species but rather belong to previous configurations of the current living world. Species and subspecies extinctions have mainly, if not only, resulted from the behaviors and acts of members of the human lineage. A diagram of the proposed origin and evolution of eukaryotic lineages is shown in Figure 8.

What then is the mechanism that has driven the eukaryotes of the early biosphere to their respective modern phenotypes? What is the mechanism of the parallel evolution of eukaryotes?

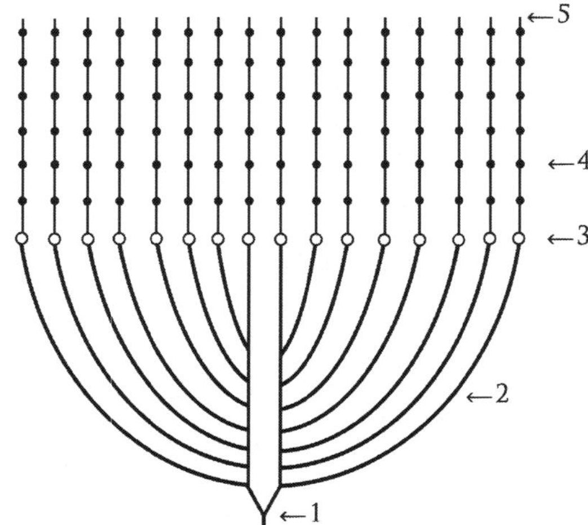

Figure 8: Diagram of the origin (1-3) and evolution of eukaryotic lineages.*

According to the seeds-first theory, each eukaryotic lineage arose from the zygote, or seed, produced by the fertilization of a gammaproteobacterium by a strain of a retroviral quasispecies. In the case of the animal kingdom, this bacterium is *Thiomargarita namibiensis.*

1- Prototype of a group of viral ancestors of eukaryotic lineages.
2- A strain of the retroviral quasispecies.
3- Gammaproteobacterium. (Final fusion leading to the Precambrian appearance of animals, for instance).
4- When a lineage has changed configuration enough to pass from one species to another. (Punctuated equilibrium).
5- A eukaryotic lineage.
* (1) and (2) make a viral quasispecies or a portion of it. Interactions between members of the quasispecies and bacterial hosts that precede the final fusion (3) are not shown. The arborization pattern of the quasispecies as represented here is only for illustration and is not necessarily indicative of the historic patterns that led to eukaryotes.

As we have explained before, a phenotype is largely the expression of a genome. Any changes in phenotypes must necessarily be preceded by some form of structural or functional genomic modification. And genomic changes, to have any consequence for the future of a species, must take place within the germline cells of a generation that obviously is not impacted by this innovation. We contend that the evolution of eukaryotes results from non-meiotic genomic and epigenetic modifications. Apprehended by us as linear changes in the fossil record, it is first and foremost an informational upgrade of the germline cells, progenitors of those organisms. We have previously demonstrated the inadequacy of the neo-Darwinian proposition that point DNA nucleotide mutations were the providers of the "raw material" for evolution. The mechanism of change through horizontal gene transfer used by viruses and bacteria also does not seem to be prominent in eukaryotes. Indeed, notwithstanding the presumed invasion of the koala genome by the retrovirus KoRV,[50, 51] viruses are not generally known to reproduce through infection of eukaryotic gametes. In practical terms, how does a reptile transform irreversibly into a bird? Besides point mutations and horizontal gene transfer, what are then the mechanisms of change within a genome?

We have agreed that the process must start as changes within the germ cells of previous configurations that probably accumulated and remained unexpressed for a while. The coordinated expression of new sets of genetic and epigenetic information has the potential to exert drastic phenotypical changes. We propose that evolutionary transitions would require one of the following three mechanisms, change in the expression pattern of preexisting genes, the creation of new genes, or a combination of both modalities.

It is within this context that we consider genetic innovation primarily through recombination as the mechanism par excellence of eukaryotic evolution. DNA recombination during meiosis between homologous chromosomes, referred to as chromosomal crossover or preferably homologous recombination, helps to produce genetic diversity among offspring through the creation of different combinations of genes. Charles Darwin is probably one of the first scientists to refer to this phenomenon—not yet understood at the time—when he wrote in *The Origin of Species*: "But I am most strongly inclined to suspect that the most frequent cause of variability may be attributed to the male and female reproductive elements having been affected prior to the act of conception . . ." He further expounds, "Seedlings from the same fruit, and the young of the same litter, sometimes differ considerably from each other, though both the young and the parents . . . have apparently been exposed to exactly the same conditions of life."[52]

Nonetheless, while homologous DNA recombination generates diversity, it does not create new genes, nor does it operate as a gene switch or regulator of gene expression. Therefore, it does not fit the bill as the primary mechanism for rearranging genetic information before the major transitions of eukaryotic life forms. Rearrangements of segments of DNA within the germline cells by the method of transposition seem to be the modus of evolutionary changes. Germ cells, especially the male line, represent the laboratories par excellence of evolution. Sperms and eggs coevolve, further delineating the frontiers and specific attributions of each lineage. Moreover, within the class of mammals, oocytes have approximately the same phenotypical appearance while spermatozoa display significant species-specific morphologic variation.[53] Investigators have also found evidence of particular genes arising de novo in Drosophila species, whose expression is said to be testis-biased and which exhibit germ-line functions.[54-56]

Transposable Elements

Transposable elements are small genetic units able to replicate and move or jump to new locations in the genome. They constitute the supreme engine of genetic recombination. They facilitate the adaptation of an organism to new challenges over the short term but are also involved in long-term evolutionary development.[57-60] A great deal has been learned about those mobile elements since their discovery in the 1940s by Barbara McClintock while studying variations in maize phenotypes. Transposable elements or transposons are active in both prokaryotes and eukaryotes. They constitute well-known drivers of evolutionary development.

Eukaryotic TEs are grouped into two classes, respectively class I and class II transposons. Class I transposons operate via a messenger RNA intermediate, which is converted back into double-stranded DNA by the reverse transcriptase enzyme. They are said to generally use a "copy and paste" mechanism. Conversely, the DNA-mediated or class II transposons excise themselves from the genome as double-stranded DNA and reinsert elsewhere according to a "cut and paste" mechanism. Mobile elements that transpose through a DNA intermediate are often simply referred to as transposons, most particularly in the case of bacteria, while those that use an RNA intermediate are called retrotransposons.[58]

Indeed, the movement of retrotransposons is similar to the replicative modality of retroviruses, as previously alluded to. Most eukaryotic transposable elements are retrotransposons. In fact, whereas retrotransposons account for almost two-thirds of the human genome, the share of DNA-based transposons is only about 3%.[61] Overall, transposable elements in eukaryotes represent the intergenic non-protein-coding DNA, previously and erroneously for that matter referred to as junk DNA. Transposable elements, most particularly retrotransposons, are believed to play the central mechanistic role in the evolution of eukaryotes. Retrotransposons reintegrate the eukaryotic genome through the action of the reverse transcriptase enzyme, the same enzyme that played a pivotal role in the emergence of the domain Eukaryota. The reverse transcriptase enzyme has then guided both the emergence and the evolution of eukaryotes. It is within this context that we also consider the evolutionary mechanism of eukaryotes as enzymatic in nature.

TEs modify the eukaryotic genome through at least two major mechanisms: Modification of the expression pattern of preexisting genes, and the creation of new genes. To transform a reptile into a bird, or *Australopithecus* into *Homo habilis*, some of the genes of configuration A will have to be expressed differently in configuration B, and it may take the creation of new genes.

Modalities of TE-Induced Evolution of Eukaryotes

A- Modification of the Expression Pattern of Genes

We propose that variations in gene expression patterns constitute one of the major mechanisms that underlie the linear evolution of eukaryotes. The current framework of evolution being that of the divergence of the species instead of a straight-line pattern, studies looking into the possible mechanisms of the phenomenon have, for instance, compared the expression pattern of similar sets of genes between related lineages. While our model is different, we will, however, use the findings of those studies to buttress the contention that the differences between successive configurations of a lineage can be also explained in part by differential patterns of gene expression. The idea that gene regulation can play a decisive role in phenotypic evolution is not new. Roy Britten and Eric Davidson have suggested, in the second part of the 20th century, the involvement of the intergenic regions of DNA in gene regulation and proposed that they play a significant role in phenotypic diversity.[62, 63] In 1975, King and Wilson argued that variations in gene expression may account for the phenotypic differences between closely related lineages, such as between humans and chimpanzees.[64]

A comparative analysis of the gene expression level in the dorsolateral prefrontal cortex (DLPFC) of humans and other primates reveals a delay in the ontogenetic development of the human brain as compared to other primates.[65] This relative human developmental retardation or neoteny has been proposed as a possible explanation of some human-specific features such as an increase in brain size and other cognitive traits through an extended period of high neuronal plasticity. Similar sets of genes but with different timing of expression led to significant differences in morphological phenotypes and other complex cognitive traits between humans and other primates. It can comfortably be assumed, then, that differences in the expression of the same gene or gene set from one configuration to the next can constitute a mechanism of the phenotypical evolution of the lineages. Here again, several studies support the role of transposable elements as the ultimate mediators of gene expression regulation, serving as the source of novel genetic regulatory elements.[66-69] Lynch and collaborators, for instance, relay the findings of several studies showing the presence of transcription factor binding sites within TEs that can help to regulate the expression of adjacent genes.[68] Transposable elements that have originated from the retroviral ancestors of eukaryotes continue to act as regulatory factors, modulating the expression of neighboring genes and, in so doing, influencing the evolution of the lineages in this domain.

Epigenetic mechanisms are also thought to be involved in the evolution of eukaryotes as they were found to be responsible for a great deal of gene expression differences between distinct lineages of primates.[70-72] The epigenome is made of the set of potentially heritable chemical modifications to the cellular DNA and its associated proteins. It modulates the expression of genetic information by determining which genes are turned on or off. Epigenetic mechanisms fall into three main categories: DNA methylation, posttranslational histone modifications, and modulation of gene expression by noncoding RNA (microRNA [miRNA]).[73-75]

And those epigenetic changes found themselves as well under the control of transposable elements. For instance, a natural and heritable epigenetic alteration that regulates sex determination in melon (*Cucumis melo*) has been found to result from the insertion of a DNA transposon at a certain (gynoecious) locus.[76]

B- Creation of New Genes

New genes are usually created through two major modalities: gene duplication, and *de novo* creation. The actions of retrotransposons are central in both.[77]

1- *Creation of New Genes by Duplication of Preexisting Genes*

New genes have been known to arise through the duplication of preexisting genes for about a half-century now.[78] Gene duplication was the first mechanism suggested for the generation of new genes. It is a broad generic concept that may involve an entire genome, or a portion of it such as a whole gene, an exon, or even part of an exon. The mechanisms involved in gene duplication are diverse and are reviewed by many authors.[79-81] And here again, in the process of gene duplication, transposable elements leave an unmistakable footprint.[82-85]

A new gene obtained by any mechanism serves as a "raw material" for evolution and genetic innovation. In the case of gene duplication, the original gene retains its functions while one or more duplicates can take on new functions for the organism. This process is summarized as duplication followed by divergence. This is exactly the case in notothenioid fishes, the predominant Antarctic fish group, where their antifreeze glycoprotein gene arose from the neofunctionalization of a duplicate of an ancestral trypsinogen gene through recruitment of intronic sequences to form coding DNA.[86, 87]

2- *Creation of De Novo Genes*

De novo gene formation has recently been recognized as a source of new genes. *De novo* genes are defined as genes that have evolved from previously noncoding genetic sequences.[88] Like all genes, they encode "useful molecules" such as proteins or RNAs (for example, microRNAs). These genes for which no homologs could be found in related species were previously called orphan (ORFan) genes, their open reading frames (ORF) having resulted from noncoding sequences. A surprising fact that genomic sequencing has revealed is that the genome of eukaryotes is usually composed of a few genes immersed in a sea of noncoding material, also referred to as intergenic sequences. Genome expansion in eukaryotes is caused by the action of transposable elements generating intergenic non-coding DNA which, again, constitutes the raw material for the formation of new genes. Transposable elements are then involved in creating *de novo* genes either directly or indirectly, as the latter are made from noncoding DNA.

In 2006 and 2007, David Begun, Mia Levine, and collaborators published some of the first papers making the case of newly evolved genes arising *de novo* from noncoding sequences in *Drosophila* species.[54-56] The *de novo* genes described included both structural and regulatory units. No evidence was found that they could be attributed to the duplication of ancestral coding sequences. Those early reports of *de novo* genes have found them to be male-biased and linked to male reproduction. Such a pattern of a male bias was not subsequently found to be universal, however.[89] Transposable elements have also been causally linked to the origin of orphan genes in multiple lineages.[88, 90-92]

Another famous example of *de novo* gene formation is the emergence of the genes coding for antifreeze proteins (AFPs) in Arctic codfishes. It is commonly admitted that organisms evolve in tandem with their environment. Consequently, major paleoclimatic changes such as the global cooling and ice cap formation of 10–30 million years ago have accompanied the evolution of antifreeze proteins in different life forms such as plants, bacteria, and teleost fish. Said proteins inhibit the growth of ice crystals within those organisms, which are then protected from freezing damage in icy seawater.[93, 94] One of the mechanisms of this evolutionary innovation is the formation of new genes from noncoding DNA. Baalsrud and colleagues demonstrated, for instance, that codfish AFPs most likely arose *de novo* from noncoding DNA around 13–18 million years ago, which coincides with the onset of freezing temperatures in the Northern Hemisphere.[87]

Finally, Roy Britten of the California Institute of Technology has reviewed and described the extensive relationship between transposable element sequences with human coding sequences. TEs were found to have contributed extensively to the coding sequences of structural genes. Roy also provided many examples where some protein-coding sequences are made up almost entirely of mobile element sequences.[95] For instance, the respective coding sequences of syncytin-1 and syncytin-2, proteins involved in human placental morphogenesis, appear to be made from the envelope (*env*) gene of human endogenous retroviruses.[96-98] This finding is on par with reports from Nekrutenko and Li that the coding regions of two mouse genes were found to largely consist of TEs, suggesting that the insertions of TEs might create *de novo* genes.[99] It is overall a modality of evolution by the creation of new genes from noncoding regions of DNA through the direct or indirect action of transposable elements, themselves of retroviral origin.

CRISPR-Cas: When the Old Becomes New Again

While we just laid out the ways and means by which, we believe, biological lineages have evolved over the ages, we have also discovered over the past decade just how important a player humanity can be in this process, in a way even more sophisticated than mere artificial selection as has hitherto been practiced in plants and animals. The method we are alluding to is called CRISPR; and we learned it from bacteria.

Prokaryotes use ribonucleoprotein complexes encoded in their DNA to fight against the invasion of foreign nucleic acids, whether from plasmids or viruses. In 1993, Mojica and his collaborators reported for the first time such a system discovered in the genome of a bacterium.[100] The system was later called CRISPR, for "clustered regularly interspaced short palindromic repeats," and soon, CRISPR-associated (Cas) genes were found in the vicinity of CRISPR arrays in bacterial genomes.[101] In its full description, CRISPR-Cas is an adaptive immune system of prokaryotes, such as bacteria. The effector of this immune response in type II CRISPR systems is a complex consisting of a DNA endonuclease enzyme, called Cas9, and a guide RNA that displays base complementarity with the invading DNA via its CRISPR RNA (crRNA) component. Hence the name CRISPR-Cas9, which is essentially an RNA-guided DNA cutter.

Countless scientists from laboratories around the world participated in its characterization,[102] but it was not until August 2012 that the respective laboratories of Jennifer Doudna, from the University of California at Berkeley, and Emmanuelle Charpentier, then from the University of Umea, Sweden, described in detail for the first time how the CRISPR-Cas9 complex works to silence foreign DNA and how it can be re-engineered to perform targeted double-strand breaks in DNA.[103] They concluded their paper with the statement that their findings could be used for the purpose of "gene targeting and genome editing applications."[103] Similar results were soon published independently by a team led by Virginijus Siksnys of Vilnius University in Lithuania.[104]

Still without wasting time, several laboratories in addition to that of Doudna demonstrated the effectiveness of the CRISPR-Cas9 model in eukaryotic cells,[105-109] paving the way for precise gene editing by homology-directed repair in this domain of life. Jennifer

Doudna and Emmanuelle Charpentier were awarded the Nobel Prize in Chemistry in 2020 "for the development of a genome editing method."

The technology they have helped develop, in addition to its many welcome applications in therapeutics, agriculture, or basic research to name only three, can also have a distinct impact on the evolution of living organisms, especially when applied to germ cells. Finally, a sign that CRISPR technology may be consistent with previous evolutionary modalities is that, here again, the potential evolutionary mechanism is to splice a foreign genetic sequence (jumper gene) to a targeted position in the genome using a modality involving an RNA-DNA hybrid molecule as in a reverse transcriptase-integrase model used by retrotransposons and retroviruses.

PART II

On the Origin of Cells and Viruses

CHAPTER

6

The Hypothesis

> Almost all aspects of life are engineered at the molecular level, and without understanding molecules we can only have a very sketchy understanding of life itself.
>
> —Francis Crick, *What Mad Pursuit: A Personal View of Scientific Discovery*

> The relevant steps in biological processes occur ultimately at the molecular level, so a satisfactory explanation of a biological phenomenon . . . must include its molecular explanation.
>
> —Michael Behe, *Darwin's Black Box*

We have made it clear in the previous chapters that eukaryotic cells derived their cellular nature from bacteria, themselves being prokaryotic cells. Therefore, we clarify that our aim in the second part of this book is to throw some light on the origin of prokaryotes and their viruses. Much has been written in this regard, but the debates are far from being settled. Viruses and viroids, for instance, are nowhere to be found in the many versions of the tree of life, from Darwin to the latest paper on evolution. They have been and still are officially excluded from the select group referred to as "domains of life," which comprises Archaea, Bacteria, and Eukaryotes. Archaea, we recall, which have previously been included within the Bacteria group, have been classified since the works of Carl Woese as a different domain of life.[1, 2] The discovery of the nucleocytoplasmic large DNA viruses (NCLDVs) and their "virophages" has not simplified the matter, either. Here, we intend to present our contribution to this debate; except when referring to the NCLDVs, viruses will be understood according to their traditional acceptance as one of the smallest forms of life.

According to its modern definition, a virus is a nucleic acid genome packaged within a protein coat called a capsid.[3] Again, classically, viruses were defined as filterable obligatory intracellular parasites, unable to reproduce on their own and entirely dependent on the metabolism of their host cell. But that definition would include viroids as well and, by its "filterable" component, exclude the megaviruses (mimivirus particles, for instance, are not filterable through the 0.2 μm pore size filters typically used to isolate classic viruses).[*4] Viruses are often contrasted with ribocells or ribosome-containing cells, which consequently can synthesize proteins.[7] While researching the literature for current opinions on the matter of their origin, we found views in some respects similar to ours expressed by the French scientist Patrick Forterre.[8] His work on the subject is therefore acknowledged here but will not be considered in this discussion.

On the other hand, prokaryotes are single-cell microorganisms that, although containing the biosynthetic machinery dedicated

---

* The discovery of viruses has a long history and involves many scientists.[5] However, the concept formally originated from the works of the Russian botanist Dmitri Ivanovski, who used a porcelain filtering device (the Chamberland-Pasteur filter) to imply in 1892 that the cause of the tobacco mosaic disease (TMD) was a filterable component, smaller than a bacterium. The pore size of the porcelain Chamberland-Pasteur filters is 0.1 μm, small enough to remove from any fluids all bacteria that were known at the time (≥0.2 μm).[5, 6]

to the production of proteins, are, contrary to eukaryotes, deprived of membrane-bound organelles such as a nucleus or mitochondria. Organisms belonging to the domains Archaea and Bacteria represent the currently accepted prokaryotes. Like the eukaryotic cell, they are ribosome-encoding structures by contrast to viruses, which rely on them for their biosynthetic and replicative needs. In the microbiological literature, viruses of Archaea are called "archaeal" viruses, while those that infect bacteria are called "bacteriophages" or "phages" for short. One might envision grouping the viruses that infect both domains under the appellation of "prokaryophages." In that same line of thought, the term "cytophages" could be considered to regroup all viruses, most particularly those whose reproduction is associated with the lysis of their cellular hosts.

Even at the risk of entering an eternal cycle of looking for primary causes, the question of the origin of prokaryotes and their viruses represents a legitimate interrogation that arises from the claim that eukaryotic cells come from viruses and bacteria. Prokaryotes reproduce autonomously by binary fission, while viruses are obligatory intracellular parasites. The latter are completely dependent on cells for their reproduction. We can then infer that the status of virus, which is obtained essentially by the acquisition of a protein shell or capsid, could only have been imparted to them by cellular structures. Before the apparition of eukaryotes, cellular organisms such as RNA cells and later prokaryotes must therefore have been their only such hosts. It is then conceivable that certain viruses might have originated from some viroids that used RNA cells and later prokaryotes for replication, acquiring in the process a capsid and the status of viruses. This scheme would explain the origin of RNA bacteriophages, but what about the great majority of bacteriophages that have a DNA genome?

First Clue: Comparing the Genetic Material of Viruses vs. Prokaryotes

Viruses are remarkably diverse, and the consensus is that many members of the virosphere are still waiting to be discovered. They can have either an RNA or a DNA genome. The host range of a virus is very narrow (one or few hosts) and most often a particular virus may choose to replicate within specific cells of a multicellular host. On close examination and as we have previously implied, it does not seem that RNA and DNA phages originated in a similar fashion. We are positing that bacterial viruses emerged in more than one way. For instance, RNA bacteriophages likely originated from naked RNA structures that parasitized cellular hosts for their replication and acquired viral status from this interaction. We are also submitting that some viruses and their prokaryotic hosts diverged from RNA cells that, according to their type, were partitioned either into bacteria and bacteriophages or archaea and archaeal viruses. This type of virogenesis and prokaryogenesis will be the focus of this part of the book. The genetic material of bacteriophages and their mode of reproduction may give us the first clues as to their origin. The genetic information of prokaryotes is encoded in DNA, while for viruses, it can be either in RNA or DNA. Chronologically, RNA is considered the first molecule of heredity, and RNA cells are the first cells to have appeared.

The evolution from RNA to DNA as a repository of genetic information can be considered an upgrade of the heredity molecule. DNA is a better compound than RNA to store genetic information. For one, contrary to the very error-prone RNA-dependent polymerases involved in the replication of the RNA molecule, DNA-dependent polymerases incorporate a proofreading mechanism, as previously reviewed. DNA has another advantage over RNA in that it has the base excision repair (BER) tool that can fix the occasional mutation caused by deamination of cytosine to uracil. One more handicap that can affect the stability of RNA is the high susceptibility to hydrolysis of the 2'-hydroxyl group on the pentose ring.

However, a certain number of events had to happen within the RNA cells before they could make that partition leap, which is also associated with the advent of DNA. For instance, the pathway allowing the synthesis of the base thymidine through the methylation of uracil had to be in place. The RNA base uracil (U) is replaced by thymidine (T) in DNA. What had to be ready as well was the process needed to synthesize deoxyribonucleotides from ribonucleotides and that involves the reduction of the ribose sugar into deoxyribose. Indeed, in DNA cells, the deoxyribonucleotide triphosphates (dNTPs) associated with the respective four bases of DNA are still produced by the reduction of ribonucleotides di- or triphosphate under the catalytic action of ribonucleotide reductases.[9] Last but not least, it was imperative for the reverse transcriptase enzyme (a DNA polymerase) needed to synthesize DNA from RNA to have been available.

Partition of the RNA Cells

The theory of the partition of the RNA cells contends that the latter will undergo a process resulting in their asymmetric division into two different biological entities: an encapsidated genetic material (in fact a retrovirus) and a cellular remnant, in most cases a precursor of prokaryotes.* This partitioning process happened in a manner akin to the eukaryotic post-transcriptional processing of pre-mRNA, where the introns are cleaved out and the exons ligated to form a mature mRNA. It comprises the induction of the formation of a

* Whether the partition of an RNA cell produced more than one retrovirus is not known. Our use of the singular in this context when talking about the number of bacteriophages resulting from the asymmetric division of an RNA cell does not negate this very possibility.

capsid by some of the formerly non-expressed genetic sequences of the RNA cell genome. Most of those sequences are then subsequently cleaved and cyclized into a lariat structure. This genetic lariat loop, along with its RT enzyme, will be packaged by the capsid, forming a nucleocapsid, a process not much different from the packaging of a double-stranded phage DNA into its prohead.[10] This structure will then bud out of the cell as a retrovirus. The cleavage of RNA sequences has been made by the breaking of phosphodiester bonds, followed by the splicing/ligation of the remaining RNA sequences.

The remaining cellular RNA material, most of which is protein-coding, will undergo reverse transcription and be transformed into DNA, forming a DNA-RNA hybrid. The RNA template will then be digested by the ribonuclease H as in a classic retroviral RNA reverse transcription scheme. The net result of this process is the partition of the RNA cell into a DNA cell and a specific retrovirus: a retrophage in the case of an RNA cell, ancestor of bacteria, an archaeal retrovirus in the case of an ancestor of archaea. It also constituted the momentous birth of DNA as the storage form of the genetic material, a critical transition in the evolution of life. The new DNA cells, precursors of prokaryotes, will multiply through binary fission. The retroviruses will, on the other hand, when they need to reproduce, return to the respective cellular matrices of their origin, now DNA cells. They will over time generate most of the DNA viruses of the modern world.

But what are nucleic acids, the structures from which stem all these developments? An account of their origin and evolution will help us further elucidate the mystery of the origin of bacteria and of their viruses.

Nucleic Acids

Despite the many differences amongst living organisms, from the simplest viroids to the most complex mammals, all life forms are based on genetic material made of nucleic acids. Those molecules are either RNA or DNA. The "nucleic" component of their name comes from the fact that they were first identified within the nucleus of eukaryotic cells.[11] Nucleic acids, which could be pictured as strings of beads, are polymers of more primary components called nucleotides, linked together by their phosphate groups. The phosphate groups of these polynucleotides are acidic. However, free nucleotides, which are anionic at physiological pH, are always associated in cells with the cation Mg^{2+}.[12, 13]

A nucleotide itself is composed of a nitrogenous base linked to a pentose sugar molecule to which up to three phosphate groups are attached (Fig. 9). The sugar of ribonucleic acid (RNA) is a ribose, while deoxyribonucleic acid (DNA) contains deoxyribose (Fig. 10).

There is a compelling body of evidence that life as we know it started as ribonucleic acid (RNA) molecules forming a system with RNA polymerization catalysts. The prevailing hypothesis is that of an RNA world, wherein the living molecule self-replicated or was its own polymerase. Up to that level of complexity, the task of polymerization, we believe, was performed by mineral surfaces, including clay minerals. The arguments in favor of RNA as an original molecule of life are diverse. Not the least are its simple template, which likely was amenable to its prebiotic synthesis by clay catalysts, its direct relationship with protein synthesis, and the catalytic property of ribozymes, to name a few.[14-17]

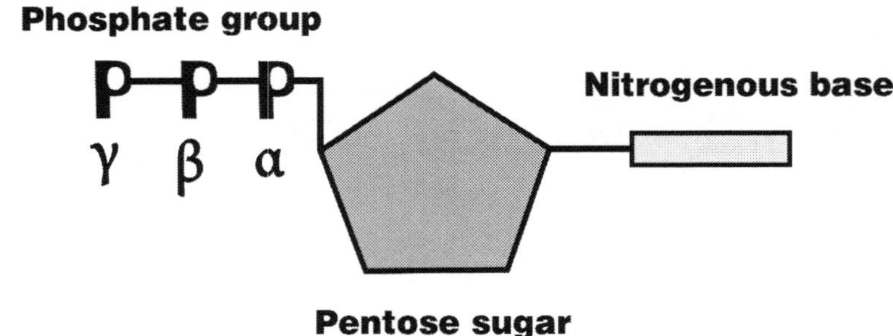

Figure 9: Nucleotide.

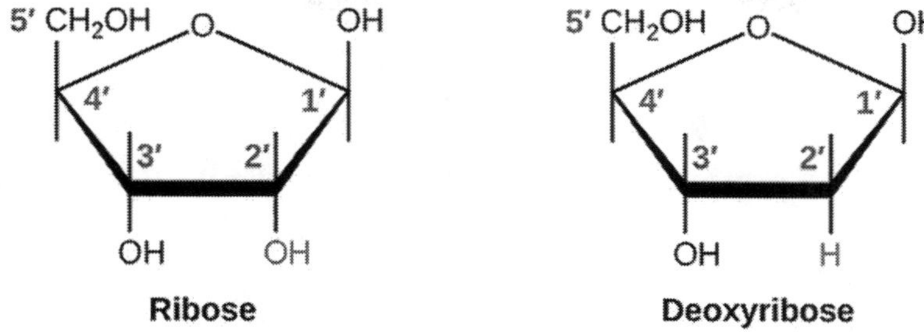

Figure 10: Pentose sugars of RNA (ribose) and DNA (deoxyribose).
[Reproduced from *Concepts of Biology, 1st Canadian Edition*, by Charles Molnar and Jane Gair, used through the terms of their Creative Commons Attribution License. https://opentextbc.ca/biology/chapter/9-1-the-structure-of-dna/]

We will assume in this endeavor that such an RNA-catalyst system came to be and started life. While we think the final say on how RNA molecules first appeared should be left to origin-of-life researchers, we will not hesitate to offer our own analysis. We will also here give a brief overview of some relevant theories and results of experiments in the field. Indeed, the chronology of events and circumstances that have led to this molecule is among the chapters of the history of life that are less well known. RNA is a polymer of ribonucleotides. A ribonucleotide comprises three essential components as previously mentioned: a base, a ribose sugar molecule, and one to three phosphate groups. The compound is called a nucleoside when it comprises only the pentose sugar and the nucleobase without phosphate groups. As for the latter, while a single phosphate group is at the center of the phosphodiester bond that links adjacent nucleosides, the level of phosphorylation of the monomeric nucleotides used as the building blocks of the RNA molecules that started life is not known.

The bases of a ribonucleotide are either purine or pyrimidine derivatives. Adenine (A) and guanine (G) are the purines while cytosine (C) and uracil (U) are the pyrimidines (Fig. 11). The nucleotides are named according to their base. When one phosphate group is attached to the carbon 3' (C3') or carbon 5' (C5'), the nucleotide is called a 3'- or a 5'- nucleotide monophosphate, respectively. Table 4 gives a synopsis of the bases found in RNA, their respective compound with ribose sugar (nucleosides), and some of their phosphorylated compounds (the nucleotide diphosphates [NDPs] are not represented).

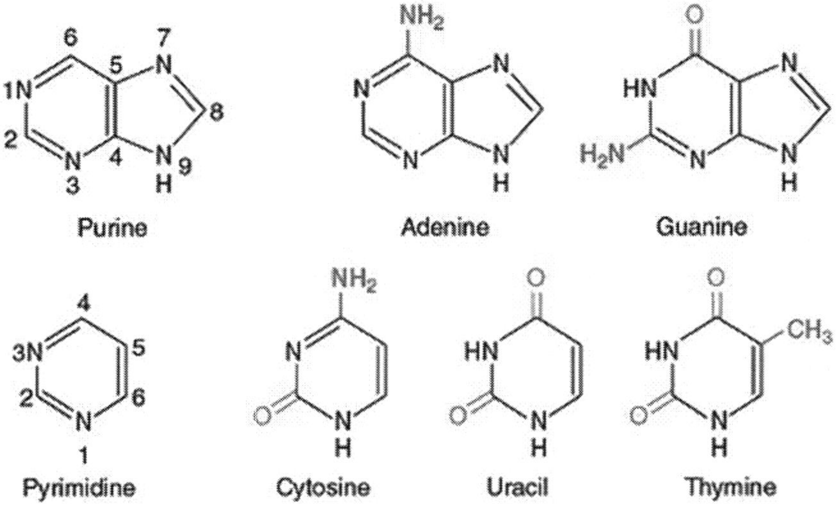

Figure 11: Basic structures of nucleic acid bases. Thymine replaces Uracil in DNA.

Attribution: http://pratclif.com/biologie-moleculaire/dna/bases.htm

Table 4: The Bases of RNA and some of their Phosphorylated Compounds

Bases	Nucleosides	Nucleotide monophosphates	Nucleotide triphosphates
Adenine (A)	Adenosine	Adenosine monophosphate (AMP)	Adenosine triphosphate (ATP)
Guanine (G)	Guanosine	Guanosine monophosphate (GMP)	Guanosine triphosphate (GTP)
Cytosine (C)	Cytidine	Cytidine monophosphate (CMP)	Cytidine triphosphate (CTP)
Uracil (U)	Uridine	Uridine monophosphate (UMP)	Uridine triphosphate (UTP)

The advent of the ribonucleotides represented a turning point in the appearance of life. They are ubiquitous molecules in biology. While they are mostly known as carriers of genetic information in the form of RNA or as part of the protein-synthetizing machine (the ribosome), they are also involved in many other processes, including energy transfer, intracellular signaling such as cyclic AMP, and more. Long before the advent of the cell and its photosynthetic production of ATP, chemical energy that could be used by precellular biological forms was carried by ribonucleotide phosphate bonds. The ribonucleotides are polyvalent molecules that are not only structural but also function as carriers of free chemical energy. In fact, ATP—a triple-phosphated ribonucleotide that functions as a major biological energy currency—has its origin in the earliest days of precellular anabolism. A case can be made that it was present at the original synthesis of nucleic acids and, arguably, that the building blocks of RNA at the emergence of life were also ribonucleotide triphosphates.

The task to create the original RNA molecules required a certain number of steps as enumerated here, but not necessarily in the same order. First, a conducive environment is required for the different stages of the process. Energy in one or more forms is needed to power the synthetic reactions. Furthermore, the precursors—namely the nucleobases, the ribose sugar, and the phosphate donors—needed to be assembled. The process is continued with the formation of the nucleosides by covalent binding of the nucleic bases to ribose sugars, followed by their phosphorylation into monomers or feedstock molecules that will be polymerized into a viable RNA molecule,[18, 19] i.e., able to serve as a template for catalytic replication. The ribonucleosides can be phosphorylated with up to three phosphate groups, respectively forming ribonucleotide mono-, di-, and triphosphates. Within a nucleic acid chain, adjacent nucleosides are linked by a phosphate group attached to either C3' or C5'.

However, phosphorylation of nucleosides can also happen at C2', and experimental phosphorylation in presumed prebiotic circumstances has also yielded 2':3'- and 3':5'-cyclic nucleotides.[20, 21] Thus, the true form and level of phosphorylation of the nucleotides that will be polymerized into the first RNA molecules are not known for sure at this time.

Catalytic units are also understood to be involved in this process. Catalytic polymerization of the monomers would lead to a biomolecule amenable to template-based replication. It was also important that the nucleotide precursors were concentrated in a location capable of countering the dilutional challenge imposed by the purely aquatic cradle of life. This role could have been played by clays, according to most origin-of-life experts.[16] Polymerization of monomers will continue until a critical mass is reached where the molecule became viable, amenable to template-based replication. The emergence of a system comprised of viable RNA molecules and their polymerase catalysts made the transition from nonlife to life. Still, before we can show the road that led us to the theory of the partition of the RNA cells to generate viruses and prokaryotes, we need to embark on a journey further back in time to the origin of the universe itself.

CHAPTER

7

Life in Gestation: Toward the RNA Building Blocks

Anabolism, or the Origin of Life's Most Direct Precursors

A small number of elements account for the greatest proportion of the composition of living matter. For instance, the elements C, H, O, N, P, Ca, and S make up roughly 97% of the dry weight of the human body.[1] When retracing the origin of life, one can go as far back as the genesis of those elements, as far as the proverbial Big Bang, about 13.8 billion years ago. The stage was laid out for the emergence of life right at the outset of this major transition. While the origin of life can be traced back to the appearance of the first subatomic particles, such as the quarks and the electrons that appeared a fraction of a second after the Big Bang, we will choose the hydrogen atom as our earliest universal common ancestor. However, it was several hundred thousand years after that event that the continued expansion of the universe allowed it to cool enough for electrons to combine with nuclei and form the first atoms.[2] Even at that early stage, in the context of an abundance of hydrogen atoms and energy, the first hydrogen molecules (H_2) are thought to have been formed.[3] In a young world that was all physics, chemistry—which in its most basic concept is the combination of two or more atoms with subsequent rearrangement of their electrons—was born. With the gravitational collapse of an impressive collection of hydrogen atoms joined by a smaller mix of helium and lithium, the first stars were formed. Within them, successive nuclear fusion reactions engineered the apparition of elements with higher atomic numbers up to iron, the 26th element of the periodic table. Among them is carbon, an essential building block of life. Almost all the elements needed to constitute living matter have then appeared before the first supernova.* In the extremely high-energy environment of those exploding stars, new fusion reactions will take place to bring about the rest of the periodic table, all the way to uranium, the 92nd element. It is also thought that some of the molecules, such as water (H_2O), methane (CH_4), carbon dioxide (CO_2), ammonia (NH_3), nitrogen (N_2), etc., which will play an essential role in the origin of life, will form as well in the throes of the first supernovae.[3] Then, when came the time for the formation of our galaxy and planet, Earth received its share of elements, molecules, minerals, and rocks that have seeded space from the supernovae. Those constituents would play a critical role in the emergence of life.

So far, we have seen the evolution of an exclusively "physics" world with the transformation of organized cosmic energy into subatomic particles of matter. Those particles combined to form the first atoms. Further evolution led the young universe to the interface between physics and chemistry with the formation of the first molecules. As to the origin of the complex biomolecules such as nucleobases, amino acids, sugars, etc., which are more directly involved in the origin of life, it is generally accepted that they have evolved from basic chemical elements available in the prebiotic era. That is in essence the proposition that formed the basis of the very

* Explosion of a star after having burned most of its hydrogen fuel.

first theories and experiments on the origin of life, even when there is still no consensus among researchers in the field as to where those initial reactions took place.

According to the Oparin-Haldane hypothesis, gaseous molecules of Earth's primordial atmosphere reacted to form simple organic compounds under the influence of ultraviolet radiation and lightning sparks. More complex, carbon-containing compounds, up to the biomolecules that started life, would result from the interaction of simpler molecules. In 1953, Stanley Miller and Harold Urey set out to test this hypothesis, and their famous experiment produced some amino acids and other organic compounds when a mixture of simple gas including hydrogen, water vapor, ammonia and methane was subjected to electric charges. The gases used as inputs were some of the same, thought to have been formed in the early era after the Big Bang, and that have seeded space with star explosions.

This Oparin-Haldane theory has been challenged by other scientists, however, some of whom proposed that the first biomolecules were rather generated around the hydrothermal vents of the ocean floor. The interface between minerals and water would provide a locus for the concentration of organic molecules in the context of the very dilute environment of the birth of life. For instance, in his book *The Story of Earth*, Robert Hazen reports results similar to Miller's, but in conditions more mimicking deep hydrothermal volcanic zones than the atmosphere or its interface with water. In some of his NASA-sponsored experiments, he reports the production of "amino acids, lipids, and other bio-building blocks" when he and his team submitted mixtures of simple gases to heat and chemically reactive mineral surfaces. He further contends that those results, which have been duplicated in numerous laboratories around the world, testify to the fact that carbon- and nitrogen-containing volcanic gases can react with rocks and seawater to produce life's basic molecules.[4] At this juncture, without taking sides in this debate, we will simply acknowledge that these complex molecules were synthesized during this time in the universe and were available on earth. Anabolism, or the synthesis of complex biological molecules from simpler components, storing energy in the process, preceded the emergence of life as we know it.

One of the first suggested cradles for the emergence of life was Darwin's conception of "some warm little pond" where primordial living forms would have resulted from the action of various energy sources upon organic and mineral precursors. This scenario envisioned by the author of *Origin* was expressed in a letter to Hooker[5] written in 1871, in which he made the point that such a scenario could not be repeated undisturbed after life has taken hold on Earth due to the dynamics of the food pyramid, characteristically unfavorable to simpler forms of life. Now almost two centuries later, currently suggested theories, while offering a much less cozy nest, have not moved too far from his aquatic cradle scenario, whether it is the dark environment of deep-sea hydrothermal vents or the wet-dry cycles of land-based hot springs. For instance, Da Silva and colleagues demonstrated the possibility of a scenario where the synthesis of RNA-like polymers could have been driven by the hydration-dehydration cycles of land-based hot springs associated with volcanic activity on the early Earth. Such natural environments are still present today, whether in Hawaii, Iceland, northern California, or in the Kamchatka hot springs of the Russian Far East.[6]

Origin of RNA, an Essential Component of Life Pioneers

The origin of the substrates that led to the first RNA molecules is still an open debate. However, several propositions fill the current origin-of-life literature. Current hypotheses and available experiments again suggest that they originated from the chemical evolution on Earth of more basic organic compounds (for example, the formamide pathway), and some may even have been delivered by meteorites during the early periods of Earth's history.

Meteorite Pathway

Indeed, a great deal of evidence has shown that a variety of organic compounds, including purine and pyrimidine nucleobases, could have been delivered to Earth by carbonaceous meteorites. The objects studied included the Murchinson, Murray, and Orgueil meteorites.[7] Galvin et al., for instance, reported that they have identified the purines adenine, hypoxanthine, and xanthine, as well as the pyrimidine uracil from the Murchinson meteorite.[8]

Formamide Pathway

Another pathway is the formamide (CH_3NO) scenario which appears to provide a route to the building blocks of RNA from the synthesis of nucleic acid bases and pentose sugars, followed by their covalent bond into nucleosides and the phosphorylation of the latter into nucleotides.

Nucleobases

Multiple studies present the evidence of the formamide origin of nucleobases under catalytic conditions.[9-13] Formamide results from the condensation of hydrogen cyanide (HCN) and water (H_2O);[14] its four elements (C, H, O, N) make it a very suitable molecule to kickstart life. As reported by Saladino and collaborators, by 2006 all the nucleobases involved in the composition of nucleic acids except guanine were part of a panel of products obtained from formamide in the presence of mineral phosphates.[15] The guanine impasse was overcome by the end of 2010 when Barks and co-workers reported the obtention of purine bases, including guanine, by heating and UV-irradiating a formamide solution.[16]

Sugar and Nucleosides

One possible pathway for the synthesis of sugars was provided by the experiment of Saladino and co-workers when they exposed formamide to proton irradiation in the presence of meteorites.[11] Different monosaccharides were obtained, including the pentoses ribose and 2'deoxyribose. The output also comprised a wide range of compounds present in living organisms, including nucleosides, amino acids, and carboxylic acids. These results not only support the possible catalytic role of meteorites but also highlight the role of formamide as parent to a wide array of life molecules, and that of radiation as a potential energy source for the abiotic synthesis of RNA precursors. The nucleosides, individually represented by a pentose sugar linked covalently to a nucleic acid base, are important in the process of chemical evolution toward life. The investigators cautioned, however, that the synthesis of those compounds by the irradiation of formamide was not as efficient as that of the nucleobases.

Phosphorylation

For some time, the question of the phosphorylation of nucleosides into nucleotides has been one of the outstanding hurdles of origin-of-life theories. For one, common phosphate minerals show limited solubility in water.[14] However, an early formamide-based chemistry has provided a possible solution to the phosphate water-solubility impasse. Indeed, as reviewed by Sponer et al., phosphate minerals are soluble in formamide in the form of metaphosphate and readily react with nucleosides.[14] For instance, Costanzo and co-workers[17] have demonstrated that at high temperature and in the presence of a high concentration of formamide, crystal phosphate minerals also at high concentration may act as phosphate donors to the nucleoside adenosine, yielding open phosphorylated forms at 2', 3', and 5' positions and 2': 3' cyclic adenosine monophosphate (AMP). Several investigators[14, 15, 17, 18] think that the cyclic nucleotides, which display more stability than the open forms, could serve as monomers or "feedstock molecules" for the abiotic formation of RNA oligomers, illustrating a possible transition from a formamide-based to a water-based chemistry. One study by Da Silva et al.[6] suggests that in simulated conditions of salty hydrothermal pools undergoing cycles of hydration and dehydration, RNA-like polymers can be synthesized non-enzymatically from 5'-phosphate mononucleosides. In summary, while we believe that the potential pathways and outcomes of the prebiotic phosphorylation of nucleosides may not have been exhausted, the observation that several phosphate minerals can efficiently phosphorylate nucleosides to form nucleotides provides a plausible scenario of chemical evolution toward life as we know it.

However, RNA is a complex molecule, and it is unlikely that it was formed by noncatalytic means. Alongside the issue of the availability of the substrates also lie those of the energetic sources and the catalytic means necessary for the prebiotic synthesis of a molecule such as RNA. Minerals and metal ions, two of the catalysts available in the prebiotic era, were essential for the formation of complex organic structures. Ferris highlights their importance in the emergence of life on our planet until a catalyst specific for template-directed RNA synthesis would appear.[19]

CHAPTER

8

Life in Gestation: Catalysis and Energy

Catalysis

The great majority of intracellular reactions are catalyzed by proteins. But, according to current literature, what catalytic systems effected the prebiotic processes?

Meteorites

We have reviewed in the preceding chapter the catalytic properties of meteorites in prebiotic synthesis. Published data show that carbonaceous meteorites could have acted as catalysts in the prebiotic synthesis of a wide array of biomolecules. Among them are the nucleobases, pentose sugars, and amino acids, in addition to ribo- and deoxyribonucleosides, etc.[1-3] Lucas Rotelli and collaborators report the obtention of some of those molecules from a mixture of formamide and water at 140° C under the catalytic action of carbonaceous chondrite meteorites.[1] In a study led by Saladino, however, a panel of life-relevant molecules was obtained when, in the presence of powdered meteorites, liquid formamide was irradiated with high-energy proton beams.[3]

Metal Ions

Metal ions represent one type of catalyst available in the primitive Earth that several prebiotic reactions could have exploited. The catalytic role of metal ions in prebiotic time can be inferred from several observations in modern metabolism. Belmonte and Mansy do not say any less when they suggest that modern metalloproteins may have evolved from prebiotic metal catalysts.[4] Magnesium ion is an important cofactor for the activity of enzymes involved in many reactions with nucleic acids and nucleotides. The flap endonucleases (FEN),[5] a class of enzymes that catalyze the hydrolysis of nucleic acids at key steps during DNA replication and repair, have an absolute requirement for divalent metal ion cofactors such as Mn^{2+}, Mg^{2+}, and Co^{2+}.

Yang and collaborators reviewed the widespread use of divalent metal ions, particularly Mg^{2+} as catalysts in the synthesis, rearrangement, and degradation of nucleic acids.[6] The role of magnesium ions in prebiotic metabolism is also supported by the observation that the self-cleavage reaction of virusoid RNAs in vitro involves Mg^{2+} or other polyvalent cations, and occurs in the complete absence of proteins.[7] Kanavarioti and colleagues also recognize the importance of magnesium ions in the template-directed synthesis of polynucleotides.[8] Furthermore, metal ions play a well-known role in stabilizing nucleic acid structures by shielding their negative charges.[9, 10] That action is further evidenced in the role played by divalent metal ions, usually Mg^{2+}, in the catalytic activity and the structural stability of ribozymes.

Mineral Surfaces

It has been speculated and demonstrated that minerals in general, and clay minerals in particular, could promote bond formation between monomers adsorbed during the repeated feedings of the wetting-drying cycles of early Earth. Benny Theng explains that by virtue of a certain number of characteristics, such as an extensive surface area and a layered structure, clay minerals were well-positioned to play a determinant role as catalysts in the chemical processes that have led to the emergence of life.[11] Clay minerals, which include montmorillonite, kaolinite, and illite among other types, are formed from the weathering of rocks and volcanic ash. Weathering in those circumstances is referred to as the disintegration of the parent structures by either a physical or chemical process.[12]

John Desmond Bernal and Victor M. Goldschmidt independently suggested early in the second part of the 20th century that clay minerals may have played a key role in prebiotic synthesis, in part due to their ability to adsorb, protect against radiation, concentrate, then polymerize extraneous organic molecules.[13, 14] Two studies led respectively by Saladino and Costanzo describe the formation of nucleic acid bases when formamide is heated in the presence of catalysts, including the clay mineral montmorillonite.[15, 16]

Besides the multiple evidence showing that some clay minerals can adsorb and concentrate key RNA components and RNA itself,[14, 17] clay-catalyzed polymerization of nucleotides into RNA has also been described.[14, 18]

James P. Ferris reports that his research group has had a better appreciation of the importance of catalysts in prebiotic synthesis while studying the formation of RNA oligomers from RNA monomers.[14] He and his collaborators have demonstrated efficient binding of RNA monomers to clay and subsequent catalysis of the formation of RNA oligomers.[18] The yield of RNA oligomers was further improved with successive manipulations on the left side of the reactions, particularly with the use of a so-called "condensing agent." Ferris also indicates that his team managed to obtain up to 40-mers RNA oligomers with further modifications and in record time. Both 2', 5'-linked RNA dimers and biologically relevant 3', 5' RNA dimers were obtained. Oligomers of up to several decamers long could be obtained after several feedings when reactions were made in the presence of mineral surfaces.[18, 19] At any rate, there is no question that mineral surfaces have played a pivotal role in the chemical processes en route to the emergence of life as we know it.

Energy

Life is an ordered state, which according to the second law of thermodynamics is not permitted without an energy flow. This law is beautifully illustrated in the popular nursery rhyme "Humpty Dumpty," which alludes to the impossibility of reconstituting a broken egg.[20] The entropy of a closed system can only increase. Life, contrariwise, is an open system. Its processes are powered predominantly by ATP, an energy currency generated by prebiotic processes, photosynthesis, or by the free energy of the electron-transport chain produced by the breakdown of metabolic fuels. Each step along the way toward the synthesis of the first RNA molecules was powered by energy. While we have not established once and for all the exact route to that end, we can try to delineate the possible energy sources used to build life. It is kind of another way to look at the Einstein equation $E = mc^2$, where a massive amount of energy can be derived from a relatively small quantity of matter. The inverse, that energy can be converted into matter, is also true. This principle has remained valid from the energy released at the Big Bang to the making of subatomic particles, atoms, simple molecules, galaxies, up to life itself. With the prebiotic synthesis of the nucleotide triphosphates, including ATP, the form of chemical energy needed to start life became available. But what sources of energy powered the emergence of those biomolecules?

Electrical

The classic Miller-Urey experiment assumed that electric discharges in the form of lightning supplied the energy for prebiotic organic synthesis on early Earth.[21] The point can also be made that if electrical energy from lightning played a role in primordial metabolism, it is most likely to have been a quite limited one due to its tendency to destroy large molecules, injure rather than build life systems.[22]

Mechanical

While a meteoritic collision with Earth in modern times would be deleterious to life up to a certain radius of the impact, it appears that the impactors had a role in impregnating our planet with life precursors when it was still sterile. Meteorites seem to have contributed to the emergence of life either by direct delivery of intact exogenous organic matter or by organic synthesis driven by impact shocks.[23] Here, our analysis of the contribution of mechanical energy to the advent of life is concerned with the latter modality. The meteoritic impacts on the proto-Earth of the late heavy bombardment are believed, for instance, to have harnessed huge amounts of kinetic energy that were converted into massive amounts of heat, triggering the chemical transformation of formamide into nucleic acid bases.[24]

Thermal

The conversion of thermal energy to a usable form is not the norm in modern biology. Nevertheless, due to its ubiquity in early Earth, one can argue that it was a reasonable form of energy to harness. It is a well-known principle that for free energy to be harnessed from heat, a thermal gradient is required.[25] Many environments, such as volcanic hot springs, tidal ponds, and deep-sea hydrothermal vents, particularly ubiquitous in early Earth, provided such a gradient.

As previously stated, current models of prebiotic organic synthesis illustrate the possible role of heat energy in the emergence of life. The involvement of thermal energy in the origin of life has been shown at least in the formamide pathway to lead to the production of nucleobases, the phosphorylation of nucleosides, and the formation of RNA-like polymers.[16, 26, 27] In this instance, the polymerization of mononucleotides into an RNA molecule occurred by the synthesis of intermonomeric phosphodiester bonds during the hydration-dehydration cycles of hydrothermal pools.[27]

Chemical

The free energy released in the course of certain inorganic redox reactions is exploited by some of the most ancient organisms. Those organisms, called extremophiles due to their living in extreme environmental conditions of temperature, ph, or chemical concentration, can use chemical energy independently from photosynthesis and aerobic respiration.[22] For their special respiration, different molecules interact to generate the movement of electrons. Examples of electron donors include molecular hydrogen and hydrogen sulfide, while carbon dioxide, oxidized sulfur compounds, and nitrate ion, for instance, can serve as electron acceptors.[28, 29] Whether or not this modality of free chemical energy production was coupled with primordial anabolism is another question. It remains true, however, that some research findings indicate that ATP or a closely related nucleotide was an original energy currency on early Earth.[30]

RNA, the first molecule of heredity, being made up of nucleotides linked together by phosphate bonds which also carry most of the energy powering modern metabolism, it is therefore reasonable to suggest that energy accumulated during the prebiotic era in molecules destined to become informational monomers, an arrangement primarily ensuring that both energy and genetics were carried by a single chemical frame.

Electromagnetic Radiation

Photoenergy of solar electromagnetic radiation is by and large the ultimate source of the chemical energy used in modern biochemistry. Some investigators have tried to establish a direct link between solar radiation and a possible prebiotic synthesis of ATP, the principal energy currency in modern metabolism. The nucleobases have strong UV absorbance.[31] Ponnamperuna and colleagues have suggested that the absorption of ultraviolet photons by purine and pyrimidine bases could have provided the bond energy needed for the synthesis of nucleotides.[31] They reported the abiogenic and nonenzymatic production of nucleotides, including ATP, when a mixture of adenine, ribose, and phosphate was exposed to ultraviolet light. Experiments like the one led by Barks show that UV radiation could have participated in the prebiotic formation of nucleobases.[32] In addition to nucleobases, Saladino and colleagues obtained a wide variety of life-relevant compounds, including amino acids, carboxylic acids, sugars, and some nucleosides, when proton beams, mimicking solar energetic particles, were used to irradiate formamide.[3]

Again, solar energy is transformed and stored as chemical energy in increasingly complex organic compounds. Anabolism, or the synthesis of complex biological molecules from simpler components, preceded the emergence of life as we know it. Whatever will be the definitive answer to the energy question, chemical free energy was in the end generated[22, 33, 34] and carried by more complex biomolecules. The free energy that was harnessed and stored in the chemical bonds of those biomolecules could be used later by living systems to maintain their complexity and perform life-related work. Most of that energy will dissipate into the surroundings in a disordered form, thereby increasing the entropy of the universe.

In summary, the chemical compounds on which free energy acted to create the RNA monomers originated from a progressive energy buildup since the Big Bang. The point being made here is that the informational molecules at the origin of life were also energy carriers. This fact remains consistent with the first law of thermodynamics, which has for its corollary that energy is convertible. Free energy is built up progressively into a submicroscopic informational molecule, which along with a replicative polymerase heralded the phenomenon of life. The continuous processing of more and more energy by the first living structures enabled them to evolve into more complex organisms, from submicroscopic to metazoans. Free energy is the ultimate architect of biological complexity.

CHAPTER

9

Life in Gestation: The First "Breath" of Matter

Lamb: . . . Therefore, in no way could I interfere with his drink.
Wolf: You are doing just that! And I am aware that you slandered me last year.
Lamb: How could I have done so when I was not even born? I am still nursing at my mother's breasts.
Wolf: If it wasn't you, then it was your brother.
Lamb: I don't have one.
Wolf: (*Enough!*) It was then one of your loved ones.

—Jean de La Fontaine, *Le loup et l'agneau*
(Adapted. "*Enough!*" is my own)

The Monomers of the First Genomes

Uncovering the secrets of life is a long-term endeavor. We still have not established once and for all the type of nucleotides—cyclic or acyclic, activated or not—that served as building blocks for the first genomes. The level of phosphorylation has not been established either. It is also a widely accepted concept that the transition from inert chemicals to living systems required the buildup of compounds relatively rich in energy, given that life itself is a highly ordered, low entropy state. With RNA, the first genetic molecule, being made of a string of nucleotides including ATP, it stands to reason to wonder whether those informational monomers would have also acted as sources of chemical energy for their polymerization. The polymerization of the nucleic acid monomers either required an external energy source or was an exergonic reaction between the substrates. In the latter case, while the synthesis of the nucleotides required free energy from an outside source, their polymerization would have been spontaneous, i.e., without the input of additional energy, but still requiring the necessary catalysis. Ribonucleotide monomers linked together by phosphodiester bonds form the building blocks of genetic molecules. Living systems endowed with metabolism use some of the same monomers in the form of nucleotide triphosphates (NTPs) to power their reactions. It bears repeating the question of whether those same triphosphated nucleotides were present from the beginning, acting as both feedstock molecules and energy sources.

Ferris's Activated Monomers

James P. Ferris has described experiments of clay-catalyzed polymerization toward the synthesis of RNA, using acyclic "activated monomers." Those were complexes of adenosine-5'-monophosphate with either imidazole (ImpA) or 1-methyladenine.[1, 2] However, it is not known whether those activation groups and their complexes with nucleotides existed in sufficient amount on prebiotic Earth. There remain legitimate questions regarding the prebiotic synthesis of activated nucleotides such as those used by Ferris in his experiments.

Cyclic Nucleotides as the First RNA Monomers

As stated previously, there is support among some scientists for the notion that the RNA monomers of the first genomes were cyclic, namely 2', 3'-cyclic nucleotides.[5-8] Those monomers can polymerize into oligomers and are presented as the most photostable among some pyrimidine nucleosides and nucleotides that were submitted to UV-irradiation, a major advantage in the ozone-layer-free prebiotic Earth.[5] They can also polymerize in the absence of templates and without the need for outside energy. As described by Dibrova and collaborators, the polymerization reaction with those cyclic nucleotides is a self-sufficient transesterification process, where the energy resulting from the breaking of one of the two phosphoester bonds is utilized for the ligation of adjacent nucleotides.[5] Indeed, an apparent support for the consideration of the 2', 3'-cyclic nucleotides as the monomers of the first genomes lies in the fact that some small satellite RNAs of nepoviruses (S-ArMV, S-CYMV, S-TobRV)* harbor a free hydroxyl group at the 5' end and a 2', 3'-cyclic phosphodiester at the 3' end.[9] However, one cannot rule out the possibility that those differences could be the result of post-transcriptional processing, such as self-cleavage, for instance.[10]

Nucleotide Triphosphates as the First RNA Monomers

That the prebiotic phosphorylation of nucleosides went as far as the synthesis of adenosine diphosphate (ADP), this is indeed one of the implications of photosynthesis. As a matter of fact, the rotary ATP synthase of photosynthetic membranes powered by the proton-motive force converts adenosine diphosphate into ATP. The photosynthetic production of ATP implies that phosphorylation of nucleosides in precellular time did not stop at the nucleotide monophosphate level, and that at least one nucleotide diphosphate

---

* **S-ArMV**: Small satellite RNA of arabis mosaic nepovirus; **S-CYMV**: Small satellite RNA of chicory yellow mottle nepovirus; **S-TobRV**: Small satellite RNA of tobacco ringspot nepovirus.

(NDP), ADP, was produced. A similar claim can be made for guanosine triphosphate (GTP) and the ribosomal synthesis of proteins, as the former plays a central role in the latter. GTP is involved in the process of protein synthesis in eukaryotes, particularly in chain elongation and translation termination.[11] In prokaryotes, as it was studied in *E. coli* bacteria, GTP is also involved in some major steps of protein synthesis, such as translation initiation, chain elongation and translation termination.[12] Moreover, histidine, an amino acid, termed essential at least for humans, is synthesized partially from ATP.[13] In fact, one of histidine's six carbons originates from ATP, one more detail that supports a pioneering role for ATP in the processes of life.

Elsewhere,[14] Vaghefi reviews the work of Shimazu and his coworkers, who demonstrated the formation of pyrophosphate bonds in an aqueous medium with phosphoimidazolide and adenosine 5'-monophosphate as substrates. The major products in the reaction mixtures were adenosine di- and triphosphates. Their study incidentally confirmed the importance of the catalytic action of divalent cations, particularly Mg^{2+}, in the formation of pyrophosphate bonds. The reactions only yielded trace amounts of NTPs in their absence.

We have also mentioned the study conducted by Chi and Kemp supporting the concept that ATP or a closely associated nucleotide may have played an important role in the processes leading to the emergence of life.[15] In the same vein, we reported the work of Ponnamperuma et al. where they demonstrated the possible synthesis of ATP in prebiotic conditions.[16] With the caveat that their experiments may not have reproduced the exact conditions of the corresponding period of Earth, they showed a possible path for the availability of ATP to the endergonic anabolic reactions leading to life. They submitted that the capture of free energy from the sun in the form of ultraviolet (UV) photons by purine and pyrimidine bases provided the bond energy for the synthesis of nucleoside phosphates; and in the particular case of adenine, we submit that this process was a precursor of photosynthesis. The absorption of UV light by nucleic acid bases is widely recognized.[17, 18] Obviously, the wavelength of light involved in photosynthesis is rather in the visible region of the spectrum and the photoreceptors are not the genetic molecule, but mainly the green pigment, chlorophyll.

Besides, almost all the known processes of the synthesis of nucleic acids use nucleotide triphosphates as monomers. In prokaryotic replication as studied in *E. coli*, deoxyribonucleotide triphosphates (dNTPs) are the monomers used as substrates to incorporate into a growing DNA chain. To that effect, cells maintain a balanced concentration of each dNTP in their stocks to minimize misincorporations or errors. DNA polymerase I, for example, one of the polymerase enzymes involved in DNA replication in prokaryotes, replaces RNA primers with dNTPs and releases pyrophosphate (PPi) in lagging strand synthesis.[19] The process is not different in eukaryotes. For instance, the synthesis of telomeric DNA by *Tetrahymena* telomerase is made by polymerization of dNTPs, namely dGTP and dTTP.[20] Deoxynucleotide triphosphates are also the monomers incorporated during the RT-directed synthesis of DNA from RNA.

Furthermore, the transcription initiation sites were investigated for the avocado sunblotch viroid (ASBVd) of both polarities in chloroplasts where it reproduces. RNA synthesis always proceeds in the 5' to 3' direction. In that way, the status of the 5' carbon of the initial nucleotide can be assessed if it was not processed any further. The process of chain growth links an incoming nucleotide through a phosphodiester bond between its 5' carbon and the 3' carbon of the growing strand. The 5' termini of primary transcripts of ASBVd(+) strands, which are in effect their initiation sites, were found to have a free triphosphate group that could be further processed in vitro with capping.[21] The investigators point out that the structural similarity between the initiation sites of both polarities of ASBVd suggests that a similar mechanism is at play in initiating their RNA sequences. Similar results showing 5' triphosphate initiation sites were found when studying the *Peach latent mosaic viroid* (PLMVd).[22] There is also evidence that the capping of the 5' triphosphate group of the chloroplast primary transcripts can be done in vitro with guanylyl transferase and GTP.[23] Besides, studies of viroid replication in vitro use NTPs as substrates instead of cyclic nucleotides, suggesting that even before the apparition of the first membranes, NTPS were used as RNA building blocks.[24]

Likewise, PCR methods for sequencing DNA fragments incorporate fluorescently labeled dNTP into a growing DNA chain.[25] Messenger RNA (mRNA) and DNA are synthesized respectively from ribo- and deoxyribonucleotide 5'-triphosphate monomers (rNTPs and dNTPs). The polymerization reaction catalyzed by specific polymerases results in the formation of a phosphoester bond between the 3' oxygen of a growing nucleic acid strand and the α-phosphate of the incoming NTP. The energetics of the polymerization reaction strongly favor chain growth, because the high energy bond between the α- and β-phosphate of NTP monomers is replaced by the lower energy phosphodiester bond between nucleotides. Besides, the energy delivered by the pyrophosphatase-catalyzed cleavage of the released PPi into two molecules of inorganic phosphate continues to drive the equilibrium of the reaction further toward chain growth.

While the controversy is far from over, all those arguments constitute ultimately a strong case in favor of the 5' NTPs at the expense of the 2', 3'-cyclic nucleotides as the RNA monomers that polymerized to form the first genomes.

Polymerization Toward Life

What is life? What is the nature of the living substance? The famed Austrian-Irish physicist Erwin Schrodinger set out to investigate this very question in his well-known book, similarly titled *What is Life?*

It may well be that the identification of the watershed event between inert matter and a living being holds the key to the answer. Life sits on the continuum of pathways and interactions that led to the polymerization of the monomers of its fundamental chemical structure. Therefore, the task of defining it is fraught with difficulties. It seems rather daunting to pinpoint a step at which the nature of the substance in question undergoes the fundamental change from inert to living. We believe that to understand the nature of life, it is necessary to identify the watershed step of that transition. The traditional view of the landmark event in the apparition of life has been the emergence of an RNA polymerase ribozyme either able to self-replicate or replicate an RNA template.[26-28] The proposition of the ability to reproduce as a necessary event at the emergence of life is consistent with our view that reproduction is a critical process, or better, an immanent feature of life as we have alluded to in the first chapter. Life as we know it is life that reproduces itself, life that creates offspring, with or without modifications. The first living beings, although by nature with a limited life expectancy, have manifested themselves through time until us, through reproduction. However, the consideration of a polymerase ribozyme as replicator in an "RNA world" scenario is a hypothesis in need of proof. Indeed, thus far not a single organism that replicates its genome with a polymerase made of nucleic acids has been found in nature.

The genome of a bacterium is replicated not by nucleic acid catalysts, but by polymerase enzymes that are proteins. Likewise, all noncellular organisms, including viroids, reproduce their genome using polymerase enzymes. As we will review in the next chapter, the reproduction of viroids, one of the simplest forms of life, involves three types of catalytic activity performed by cellular host enzymes, namely RNA polymerization followed successively by cleavage of a concatenated sequence of monomeric chains and ligation to circularize the resulting linear monomers. Viroids of the Avsunviroidae family cleave their multimeric strands into unit-length linear molecules with a hammerhead ribozyme, but they still require cellular enzymes for nucleotide polymerization. It would not be far-fetched then to assume that viroids have always used protein catalysts (which they lack) to reproduce their RNA genomes. Viroids and viruses reproduce inside cellular hosts, as the latter concentrate the necessary nucleotides, amino acids, and proteins needed for such an endeavor.

Hence, we wonder whether proteins or polypeptides* replicated the first genomes as well. The affirmative answer to this interrogation would imply the availability of life-relevant protein enzymes before the very first RNA replication. RNA and proteins would then have been practically contemporaneous molecules. The necessity of protein catalysts to reproduce the first genomes seems then to be one of the major constraints for the emergence of life. In light of the above, the stipulation of RNA acting as its own polymerase in a so-called RNA world is untenable and, as a result, we must account for the nature of the catalysts that replicated the very first molecules of heredity.

We must then compare the topic of replicability against that of self-replicability. Self-replicability is out of the question for the

---

* For the sake of simplicity, we will hereafter in this chapter use the term "proteins" for both in most instances, even though the initial enzymes could have been just polypeptides.

pioneer genetic molecules when there is no evidence of RNA acting as its own polymerase[29] nor of proteins carrying out reflexive polymerization. We can consider prokaryotes and eukaryotic cells as indisputable prototypes of self-replicability. The genetic heritage, the repository of biological information is carried in all cells by nucleic acid molecules, RNA in the case of RNA cells and DNA in modern cells. The reproduction of this biological information is catalyzed by protein enzymes. These same proteins derive from information encoded in the genetic molecule and the last step of their production is catalyzed by nucleic acids in the forms of RNA, as it happens, the ribosomal RNA (rRNA). It has been well demonstrated that it is in fact the ribosomal RNA that provides the catalytic activity responsible for the formation of peptide bonds between amino acids leading to the ribosomal synthesis of proteins (See also next chapter).[58, 59] The genetic instruction for protein synthesis reaches the ribosomes as an RNA messenger, and transfer RNA (tRNA) molecules carry amino acids to the site of protein synthesis. It follows that the catalysis of the expression or reproduction of nucleic acid molecules by proteins is coupled to the catalysis of the production of those same proteins by nucleic acid molecules. A type of circular process that the physician-biologist Stuart Kauffman had summarized in the concept of autocatalysis in his description of self-sustained networks of chemical reactions he called autocatalytic sets.[60, 61] The different elements of the cellular catalytic loop essentially catalyze the maintenance and the reproduction of the cell.

Cells not only reproduce but are also endowed with metabolism; they synthesize organic molecules, respire, eat, digest and excrete among other functions. Quite obviously, they are the products of an evolutionary process from a simpler system. We are here making the case that life did not arise as the well-formed cells we know, but rather as precellular RNA-protein systems which were also self-sustaining provided the existence of an appropriate environment containing the necessary nutrients ("food set"[61]). At the center of all living systems, we contend, was always a network of interdependent molecules whose formation has entered mutual catalysis. We could find no other way for life to have emerged when considering the imperative of reproduction as previously delineated. Life, at its minimum level of complexity, is made up of proteins and ribonucleic acid (RNA) molecules. Proteins catalyze the reproduction of nucleic acids. Nucleic acids must then catalyze the production of proteins. Living forms are essentially nucleoprotein systems.

It is not hard to realize that viroids and viruses display some features of precellular living systems. Indeed, there is something fundamental about viroids and viruses, noncellular structures that replicate using the enzymes synthesized by other systems. On par with the erroneous comparison often made between the queen and the worker bees in a hive, it is commonly believed that the genome is a passive molecule and that enzymes are the workhorses of the cells.* However, a viroid or a virus at the right moment will, from their reservoirs, find the appropriate host to trigger their reproduction, not the other way around. Human influenza A and B

* Actually, one can view a beehive as a cell where the queen bee represents the genome, while the worker bees would stand for the protein catalysts that make things happen. Or it can also be viewed as a multicellular organism with the queen bee at the place of the gametes, and the worker bees at that of all the other cells and organs. The same can also be said of social ant colonies. Life reproduces the same patterns at different levels of organization. The queen bee plays a very active role in the life of the hive. She takes the nuptial flight that culminates in the fertilization of her eggs. The fertilized eggs will then produce a continuous supply of female workers. The queen is a key player in the life of the hive, just like the genome in that of the cell.

viruses, for instance, make their reproductive rounds every year at about the same period. All this to emphasize that the picture of the genetic molecule as an idle chemical that gets passively replicated by enzymes may not be completely accurate. Therefore, something would be amiss if we summarily disqualify viroids and viruses from the world of living matter. Like them, the pioneer RNA molecules of life were not passive entities; they could not self-reproduce, but they sought out their reproduction. They formed living systems with the molecules that catalyzed their reproduction. We will then hold the emergence of autocatalytic sets of RNA molecules and proteins as the critical event (Kauffman's phase transition) in the emergence of life. The achievement of "catalytic closure" by the different elements of the catalytic loop is the watershed event heralding biosystems. Life appeared in an environment able to concentrate the monomers from which the nucleoprotein* system is constructed. In the language of complex systems, those monomers form part of the "food set."

Again, considering the reproductive modality of viroids and viruses, we affirm that the first living systems were nucleoproteins. They consisted of protein catalysts and ribonucleic acid molecules (genome, ribozymes and other ancillary molecules such as tRNA precursors) (Fig 12a). Whether mineral surfaces continued to produce the proteins of the first nucleoprotein systems under the induction of the nucleic acid genomes or that protein-producing ribozymes were part of the original systems is not cleared. But it would not be surprising, and we assume here, that mineral-synthesized proteins constituted a sort of umbilical cord linking the first nucleoprotein systems to their mineral progenitor until their own ribozymes could take over the protein production process.

---

* Nucleic acids and proteins.

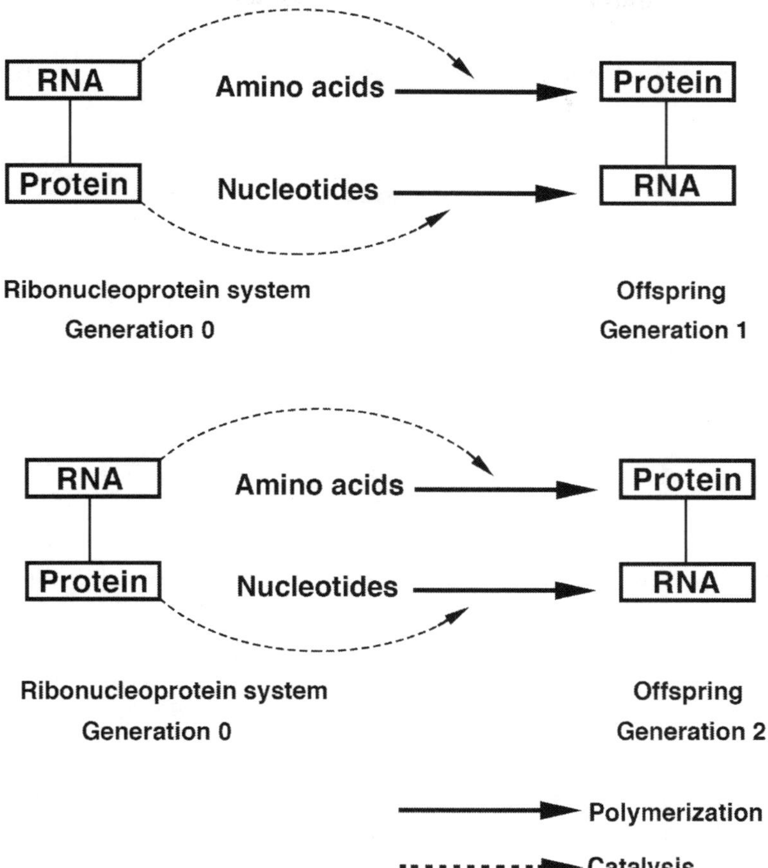

Figure 12a: The interface between chemistry and biology. Diagram showing two rounds of autocatalysis of a given nucleoprotein set.

From the above, we note the relevance of certain questions such as the following. From simple monomers, how did this planet end up with complex living systems made of RNA and proteins? Life is believed to have emerged in the general context of an aquatic macroenvironment.[30, 31] How then did the problem of dilution get solved to allow for the concentration of monomers in specific areas? What catalyzed the condensation of those molecules into long polymeric chains? The polymers intervening in an autocatalytic set only catalyze the formation of a new set, but whence cometh the original autocatalytic sets of nucleic acids and proteins?

One obvious requirement is therefore the presence of specific environments able to concentrate and preserve the prebiotic molecules en route to constitute the first living systems. The catalysis of intermonomeric bonds—phosphodiester in the case of nucleotides, and peptide in the case of amino acids—was also vital, if you will, to this endeavor. Mineral surfaces, particularly clay minerals, fulfilled both of those tasks, according to a proposition that seems to derive some support from the available evidence.[32, 33] Clay minerals also appear to offer a sanctuary to the nascent RNA polymers as well as to their parent monomeric molecules, protecting them against many environmental insults such as UV radiation.[32]

We propose that the first living entities were chemical systems that emerged in the interlayer surfaces of minerals. They consisted of RNA molecules and proteins. Structurally, life is then necessarily bifactorial. There is no life without nucleic acids. Neither can life exist without peptides or proteins. Besides the requirements of a conducive macroenvironment, the confluence of three major protagonists—namely ribonucleotides, amino acids, and mineral catalysts—brought about the living world. The last step in the emergence of life can be illustrated as a simple system model, where nucleotides and amino acids constitute the inputs, life the output, and mineral catalytic activity the process (Fig. 12b). It should be emphasized, however, that while the propositions developed here in that respect derive some support from available research, they should remain only as insights or hypotheses awaiting the ton of research that this area desperately deserves. Let us then address the questions about the process of polymerization.

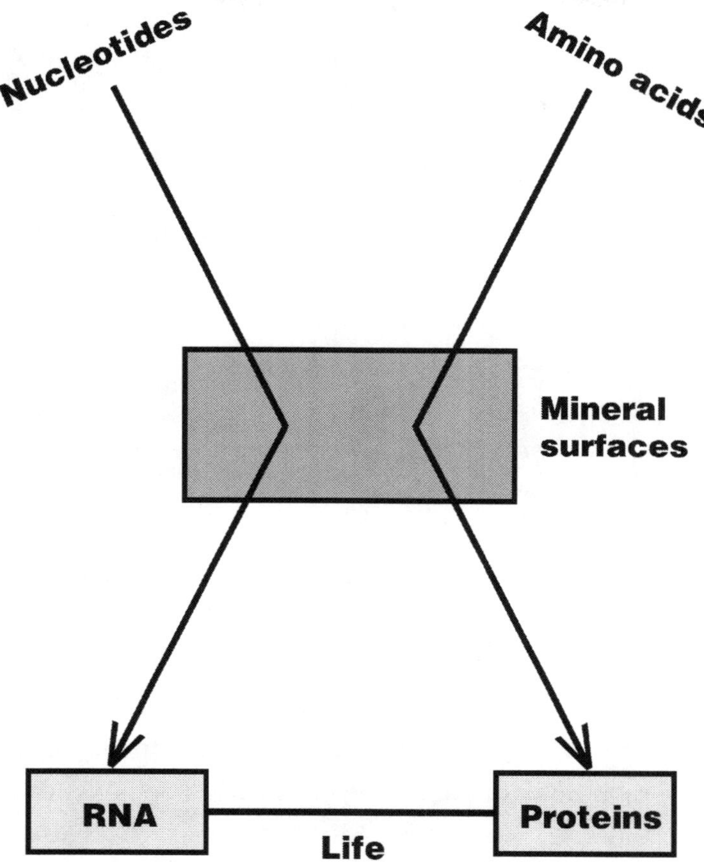

Figure 12b: The last step in the emergence of life. Mineral surfaces catalyzed the formation of the first living systems which were autocatalytic sets of nucleic acid and protein molecules. Mineral surfaces are, in essence, life factories. Overall, we believe and affirm that the origin of biological information and of its translation code, namely the genetic code, lies in the structure of minerals.

Mineral Surfaces as Bond-Makers and More

In the context of origin-of-life research, clay is the most studied of the minerals. Clay, we recall, is a sedimentary material that originates from the weathering of rocks and volcanic ash. Clay minerals are organized into negatively charged layers that are retained together by secondary chemical bonds and interlayer cations.[2, 34] Their net negative charge is proportional to the degree of isomorphous substitution, for instance, of Si^{4+} in the tetrahedral sheets or of Al^{3+} in the octahedral sheets by lower-valent cations. The layers themselves are made of 2–4 sheets linked together by primary (including covalent) bonds, which are generally much stronger than the secondary ones. Therefore, the weakest point of clay is at the level of the interlayer bonds.

Montmorillonite is one of the easiest to split clay minerals due to the weaker interlayer bonds, made of Van der Waals forces instead of the stronger hydrogen bonds found in kaolinite, for instance. Alexander Graham Cairns-Smith has suggested that clay minerals could self-replicate by cleaving or splitting, followed by the growth of the pieces. The crystals-as-genes model that originated from his publications proposes that clay crystal growth imperfections, the distribution of charges, or the aperiodic stacking of their layers constitute information or "primitive genes" similar to the nucleotide sequences of DNA and which can also be transmitted to "daughter" crystals.[35] In a sense, a clay mineral-based "life" would have preceded nucleoprotein-based life (i.e., life as we know it). "Clay-life" would have then served as scaffolding and a template for the establishment of modern biochemistry through the synthesis of the first polypeptides and RNA oligomers.[35] In other words, geochemistry "handed" the code of life over to biochemistry.

We have heretofore focused more on the emergence of RNA. What then of prebiotic polypeptides? It is important here to note that amino acids, monomers in the synthesis of proteins, also fundamental constituents of life, were in principle obtained by the Miller experiment.[36] Luca Rotelli and collaborators have also demonstrated the obtention of some amino acids from formamide in aqueous media under the catalytic action of meteorites.[37] Moreover, B. Theng reviewed several reports on the synthesis of organic substances, including amino acids, under the catalysis of clay minerals.[38]

However, if mineral surfaces, in general, have been shown to participate in the synthesis of amino acids and the nucleobases,[38-40] their best-proposed involvement in the emergence of life is the processing of ribonucleotides and AA into RNA and polypeptides, respectively. The mineral-catalyzed polymerization of those monomers is well established.[1, 38, 41, 42] Using successive "feedings" of mineral surfaces with the activated monomers (montmorillonite for nucleotides, illite, and hydroxylapatite for amino acids), a method that mimics the hydration-dehydration cycles of volcanic hydrothermal fields,[31, 42] Ferris, for instance, reports that he was able to obtain oligomers of several decamers long.[1]

Layered double hydroxides (LDH), another class of minerals common in early Earth, display a layered structure like clay but with positive layer charges instead, a property that allows them to concentrate deprotonated amino acids in their interlayer space and polymerize them. The charge of amino acids depends on the pH of a solution, a feature that allows their adsorption onto different minerals in different environments. A "computational modeling study of interactions between amino acids and LDHs" carried out by Erastova and collaborators shows that the minerals can swell, adsorb, and concentrate amino acids, and promote peptide bond formation during alternating hydrated and anhydrous conditions, such as what occurs with the wet-dry cycles of volcanic hot springs.[42] Those conditions are thought to have been common in prebiotic Earth.

As for energetic requirements, the polymerization process of RNA monomers advances without the input of outside energy.

The energy of the broken phosphate bond is conserved and used in the ligation of a new phosphodiester bond. Concerning the peptide bond formation on LDH surfaces, while it is an endergonic process, it is nonetheless energetically favorable.[42] Rehydration of the interlayer space, which is assured by the release of a molecule of water with each peptide bond formation, drives forward the polymerization process. Up to a certain degree of polymerization and even organization, despite the obvious presence of complex molecules, life, as it started, — a mineral-synthesized autocatalytic set of RNA molecules and proteins—, did not yet exist. A critical RNA mass (greater than 40-mers[2]) needed to be reached before an RNA molecule could serve as a template for the mineral-produced polymerase, part of the momentous transition from chemistry to biosystem, from inert to living matter.

However, besides polymerization, are there other specific contributions of mineral surfaces to the emergence of life?

It is believed that clay minerals preferentially adsorb and polymerize one optical isomer of a molecule over the other. It has been found, for instance, that montmorillonite catalysis of a racemic mixture of D and L nucleotides leads to the formation of RNA dimers that are primarily D-D or L-L,[2] instead of a mixed D-L pot. In that vein, as it relates to their structure, living systems selectively use homochiral (same handedness) biomolecules, in the sense that for each of those compounds, one stereoisomer is preferred over the other. The selective use of D-pentose sugars in RNA and DNA is associated with the synthesis of proteins almost exclusively made of L-amino acids.[2, 44-46] The discriminate adsorption, by clay minerals in particular of a specific enantiomer of those molecules is purported to be at the origin of the D- and L-homochirality respectively of nucleic acid pentose sugars and the amino acids of living organisms.[37]

Another pertinent question regarding the involvement of minerals is whether polymerization proceeds according to a preexisting instruction derived from their structure, as speculated by Cairns-Smith. At any rate, Ferris indicates that one of the observations he made from the experiments he participated in is that the sequence of nucleotides resulting from montmorillonite-catalyzed formation of RNA oligomers bears the signature of a rather non-random process. He goes on to point out a notable phenomenon he calls sequence- and regioselectivity. Eight dimers counted for 84% of the 16 possible in experiments involving equimolecular amounts of the four bases A, C, G, and U, and all 8 of them started with a purine nucleotide at their 5' end.[2-4] In other words, through a non-random polymerization process of the ribonucleotide monomers, information contained within the structure of clay minerals was transferred to the RNA strands of the first living systems. Likewise, in the previously cited study of Erastova et al.,[42] the peptide chain synthesis at the interface of layered double hydroxides (LDH) proceeds from the N-terminus to the C-terminus, as happens also in ribosomes. The growing peptide chain rests appended to the LDH surface by the C-terminus of the last amino acid attached, and the way those C-termini were arranged onto the mineral surface was found to be templated by its charging sites, namely the distribution of aluminum atoms.

Part of the work done by minerals in this context is therefore reminiscent of the ribosomal synthesis of proteins by polymerization of amino acids. A possible major difference between minerals and the ribosome is that the instructions by which amino acid polymerization takes place, although external to the latter, seem to come from the former themselves. We have seen the indication, from the experiments just cited, that the mineral-catalyzed polymerization of nucleotides and amino acids is not random, and that information contained in the clay or the LDH is transmitted to the synthesized polymers. Indeed, although minerals have well-defined unit structures, they have many sources of aperiodicity that can constitute specified information or a meaningful arrangement of symbols. In the case of clay minerals, the different types of layer stacking or more specifically the isomorphous substitution of cations could very well play this role, i.e., be the ultimate source of biological information. The basic structure of clay minerals is represented by silicate $(Si_4O_{10})^{4-}$ in the tetrahedral sheets and $Al_2(OH)_6$ or $Mg_3(OH)_6$ in the octahedral sheets.[34] Figures 12c and 12d show the layer-lattice structure of montmorillonite clay. One quickly realizes that the monotonous succession of Si^{4+}, Al^{3+} or Mg^{2+} cations in their respective sheets would constitute only pure and simple redundancy but not functional or specified information able to direct the synthesis of complex organic polymers. However, due to the phenomenon of isomorphous substitution, from time to time, Si^{4+} can be replaced in the tetrahedral sheets by lower valence cations, such as Al^{3+}, Fe^{3+} or Fe^{2+} and Al^{3+} by Mg^{2+} or Mg^{2+} by Na^+ in the octahedral sheets.[*34] Isomorphic substitution of cations in the basic unit structure of minerals, in our view, is a strong candidate for creating the kind of variance or inhomogeneity needed at the origin of biological information according to which, at a defined synthetic step, the minerals could have attached either a certain RNA triplet (codon) in the

* The cations here would constitute symbols or characters of a specific language. In our model, the arrangement of cations in a mineral sheet and the phenomenon of isomorphous substitution are not chance events as will be highlighted by the developments of the 3rd part of the book.

synthesis of RNA molecules or a corresponding amino acid or stop sign in the synthesis of a polypeptide. This type of chemical or electrical equivalence (between codons and amino acids or stop signs) for the mineral surfaces would be the origin of the genetic code. In this line of thought, we believe that there must be some property of an amino acid, probably electrical, which reciprocates a corresponding feature in the arrangement of the associated codon, and which is then responsible for the genetic code.

Nucleic acids and proteins can be compared to two different languages linked by a translation code which establishes the equivalence between their respective words, concepts and sentences in the same way that *"Bourik chajé pa kanpé"* in Haitian corresponds to "A loaded donkey does not stop" in English.* It is exactly the same message or word of wisdom that is delivered in both languages, as guaranteed by the translation code. In the case of biological information, the chemical equivalence, assumed to be electrical, was established with the birth of the twin languages from a specified sequence of characters on the surface of minerals. A sequence of symbols on the surface of minerals (the message, or the meaning) could be translated either into a codon (the nucleic acid language) or into an amino acid or a stop command in the synthesis of polypeptides (the protein language). We are simply saying that the sequence of cations as modified by isomorphic substitution within the structural units of minerals is likely to be that message.

---

* The proverb means that one should not waste time when one has a lot on one's plate. The translation code establishes an equivalence in meaning as (*bourik* = donkey; *chajé* = loaded; *pa* conveys negation as in does not; *kanpé* = stop, stay still, stand up,) in the same manner that the genetic code links codons to amino acids or stop commands.

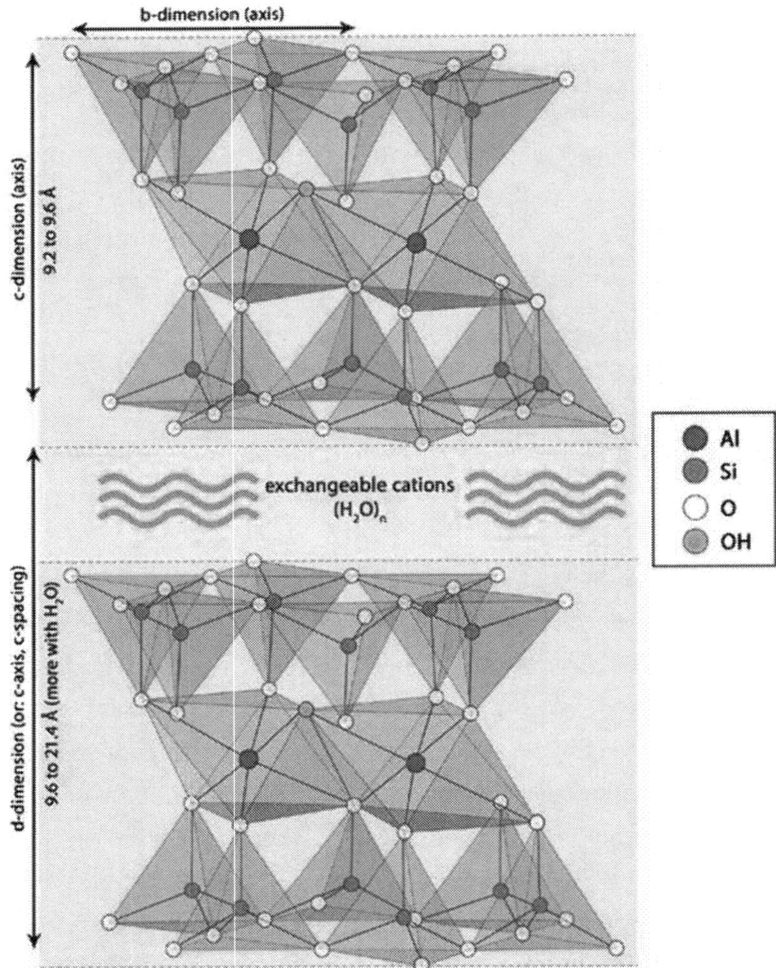

Figure 12c: Structure of montmorillonite clay.* A montmorillonite clay layer is made of two silicon tetrahedral sheets oriented toward each other while enclosing an octahedral aluminum sheet. The intersheet bonds of a given layer are strong, being of either the covalent or the ionic type. Adjacent silicon sheets between two montmorillonite layers are retained near each other by Van der Waals forces, which are much weaker. Adsorbed water and surface cations in the interlamellar space cause the material to expand.

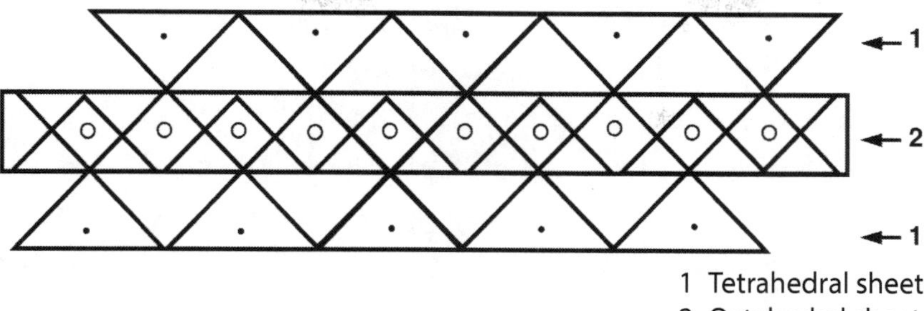

Figure 12d: A simplified drawing of a layer of montmorillonite clay showing an aluminum octahedral sheet sandwiched between two silicon tetrahedral sheets. We propose that the isomorphous substitution of some Si^{4+} and of some Al^{3+} in their respective sheets can generate the type of aperiodicity in the arrangement of cations along the mineral surface from which biological information originated.

Mineral surfaces provided, then, a path for the availability of both RNA and polypeptides in the run-up to the emergence of life. They are, if you will, the manufacturing plants from which life emerged. The extent of the role played by ribozymes and other ancillary RNA molecules in the mineral synthesis of life-relevant polypeptides, whether it would be the aminoacylation of a tRNA precursor as has been demonstrated in vitro[48] or the transport of amino acids to the site of polypeptide synthesis, etc., is not known for sure. Those important details still need to be worked out.

Another question that, immediately raised, was answered was the agency by which these early proteins acquired functional conformations beyond their primary structure. In eukaryotes, proteins, once synthesized in the ribosomes, get folded in the endoplasmic reticulum. However, prokaryotes, just like the first replicons which they follow chronologically and unlike the eukaryotic cells they precede, do not have an endoplasmic reticulum for the post-translational modification of proteins. Raimon Sabate et al. answered this question when they described several mechanisms controlling protein folding in bacteria, including genetic, transcriptional, and others that operate at the amino acid sequence level.[49]

Although clay minerals research has yielded high-quality evidence about the polymerization of ribonucleotides and amino acids, it is to be noted that a variety of metal ions has also been shown to catalyze the polymerization of activated RNA monomers.[33] Besides, as mentioned earlier, non-clay minerals such as LDH can also adsorb, concentrate, and polymerize amino acids. The abundance on early Earth of those different candidates to catalyze the polymerization of prebiotic monomers suggested the metaphorical reference to one of the most famous fables of the 17th-century French fabulist Jean de la Fontaine at the beginning of this chapter. The clay and the LDH pathways remain, however, the most plausible scenarios to account for the bridging of prebiotic monomers to the living world. Clay and LDH have opposite layer charges. LDH-montmorillonite composites have been synthesized and the resulting enhanced adsorptive capacity is used to purify water.[50-52]

Toward a Definition of Life

The proposed configuration of the first living systems is exactly that of viroids, more particularly the simplest of them, the pospiviroids: a non-protein-coding genome that replicates using protein catalysts or enzymes produced by another system. Mineral surfaces concentrated the required amino acids and synthesized the polypeptides/proteins just as plant cells have done for the viroids for millions of years. We view the first RNA genomes, like modern viroids, as forming a single system with the other structures that support them, including polymerase enzymes.

The emergence of an autocatalytic system made of RNA and proteins heralded life and any destructive process that could affect it such as resulting from disintegration or hydrolysis was its death.

Life is then an emergent phenomenon that has resulted from the organization of different prebiotic monomers into a complex system of polymers with novel properties such as "the whole is greater than the sum of its parts." As we have seen, life—like everything else in nature—obeys the second law of thermodynamics, and as such, the creation of complex living systems and the maintenance of this complexity requires free energy. Isolated, living organisms are unable to maintain their complexity and will eventually degrade due to increasing entropy. The first RNA-protein replicons faced such a conundrum. However, the confines of the protective layers of minerals have, to a certain extent, stabilized their entropy.

Nonetheless, to continue to evolve in complexity, the first replicators needed to make better use of the energy sources available on Earth. Living organisms interact with their environment from which they take free energy. This energy will be used to do useful work, while disordered energy in the form of heat will be released into the surroundings by the system. Life is then the result of a long process that started, to pick an arbitrary time frame, very shortly after the Big Bang with the creation of ordered or low-entropy states at the expense of the universe. As if the universe were not expanding fast enough to dissipate free energy into radiations with much longer waves, it resulted in the formation of subatomic particles. The process continued with the collision of those particles, their binding into atoms, the appearance of stars, and the formation of more elements. Chemistry was born from the exclusive world of physics when atoms combined to form the first molecules. In a universe where free energy was plentiful, endergonic reactions practically had no limits. Individual atoms and simple molecules interacted in the interstellar medium and on Earth, culminating in the formation of organic complexes.[33, 53]

All forms of life, from viroids to eukaryotes, contain a set of instructions made of nucleic acids, just as they all depend on enzymes for reproduction. While the evolution of matter is a continuous process, we then identify another major step toward life in the synthesis of those specific polymeric molecules, namely ribonucleic acids and polypeptides. The transformation of nucleotides and amino acids into viable polymers organized into self-sustaining systems of RNA molecules and their polymerase enzymes constitutes the ultimate and critical step in the apparition of life, the phenomenon with which biology is concerned. This was, in short, a stepwise evolution from physics to chemistry, then to biosystems. Life then rests on this original tripod of ordered state or negentropy (physics), nucleic acids and proteins (chemistry) organized into autocatalytic systems with emergent properties (transition to biosystems). It is noteworthy to mention here that given the properties of the enzymes involved in the reproduction of nucleic acids, autocatalytic systems of nucleic acids and proteins are inherently prone to generate diversity. These features constitute the common denominators of life and the only terms by which it can be defined.

The watershed step that heralded the first life forms was the achievement of "catalytic closure" within a set of mineral-synthesized proteins and ribonucleic acid molecules. The catalysts of living systems not only replicate a nucleic acid genome but also express it, for example into useful molecules and body plans. As a result of their nature, living forms multiply. Virtually all cells and multicellular organisms do. The few cells that stop dividing after reaching structural and functional maturity are nonetheless alive. They are essentially living systems that have ceased to divide in order to better fulfill a particular function. This is the case with permanent cells such as neurons or cardiac muscle cells or cardiac muscle cells which, although they do not regenerate, nevertheless produce the proteins necessary for their respective functions through the interaction of catalysts with their genes.

Some of the original nucleoprotein systems evolved into cells and the resulting organisms became either uni- or multicellular. It is here fair and proper to mention the name of Erwin Schrödinger, the famous physicist who, in the concepts of his time, correctly stated the nature of the problem in his timeless essay, *What is Life?* Furthermore, theoretical biologist Stuart Kauffman deserves immense credit for his insightful characterization of the chemistry of living forms as autocatalytic sets. We are then led to define life (as we know it) as the condition, within the limit of a decay threshold, of an entity made up of one or more autocatalytic systems based on proteins and nucleic acids. This definition constitutes the greatest common divisor of the living world.

Viroids and viruses are alive. The cellular apparatus of the cells they infect, comprising among other structures ribosomes and nucleotide polymerases, supports two living systems; one centered on the cellular genome, the other on the subcellular replicon. A multicellular organism is merely a cooperative network between individual living entities. Such an organism is said to be dead when specific patterns of cellular activity, for one reason or another, dip below a certain threshold, where its functioning as a whole can no longer be supported. However, not all the cells or even organs have necessarily died at that instant. Some of the subsystems of the said organism may take longer before decaying beyond rescue.

This is the principle used in cadaveric organ donations, a frequent example of which is the harvesting and transplantation of the cornea. The carcass of a dead animal preserved in permafrost for thousands of years may still contain undecayed DNA fragments or even cells. Again, those DNA samples, while undecayed, are not "alive" as per our definition if they are not part of an apparatus that includes ribosomes and functional proteins. While those proteins could be encoded within the DNA fragments, they need to have been expressed and be operational for the system to be considered alive. Furthermore, samples of Pleistocene permafrost deposits estimated to be about 40,000 years old by radiocarbon dating were recently thawed to reveal viable nematodes.[54] There are also examples of bacteria that have been revived and identified from samples up to 250 million years old.[55, 56] Extreme frigid conditions can suspend the vegetative functions of certain organisms while reducing to a minimum their entropy change. Likewise, hibernating animals, while they drastically reduce their metabolic requirements, live on the reserves they have accumulated and continue to breathe, albeit less frequently. The principle is not different for some gram-positive bacteria that can sporulate under harsh environmental conditions but are still alive, nonetheless. They keep ribosomes, their genomic DNA along with a minimum concentration of essential proteins within a dehydrated multilayered structure until favorable conditions return.

Additionally, we have evidence that human mature erythrocytes are RNA cells converted from DNA cells. Human erythrocytes, like mammalian red blood cells, are devoid of DNA, having lost their nucleus during the maturation process. It has been found however, that despite this apparent setback, they remain fully functional cells containing mRNA transcripts, mostly globin mRNAs, as genetic material, and ribosomes allowing them to translate those genes into polypeptide chains.[57]

While we have worked out the fingerprints of life and encapsulated its essence within the confines of a definition, we acknowledge that living beings can circumvent the laws of biology and even physics for functional reasons, occasionally by making adjustments. We postpone the effects of the second law of thermodynamics by taking in food, drinks, oxygen, etc. We escape gravity by flying or create machines that do so. Some of us avoid decay for periods of time, not by making use of a food set but by drastically reducing our metabolism (hibernation, sporulation). Viroids and viruses have found a way to keep separate different elements of their autocatalytic system. Finally, in multicellular organisms as we have mentioned earlier, some cells stop reproducing in order to accomplish very specialized functions.

The first living organisms, however, not only because they were unitary but also to create a legacy through time up to us, had to take full advantage of their autocatalytic essence and multiply. But what was their phenotype? Can we uncover their fossils?

CHAPTER

10

From Naked Replicons to RNA Cells

From its humble beginnings to its modern phenotypes, life has left along its trajectory a trail of fossils, some living, others not. The fossils we are looking for at this juncture belong to the living category, and obviously are not recoverable with the traditional tools of the paleontologist.

The Viroids

Arguably, the best-known simple RNA replicons are the viroids. A quick survey of their characteristics reveals that they are non-protein-coding and represent the earliest fossils of life (as we have alluded to in the last chapter). In reality, they are only one part of a two-part living system, the other structures including proteins that support and reproduce them being enclosed within their reproductive hosts.* Separate reviews[1, 2] led respectively by Daròs and Flores describe some of their basic features. They are circular, non-protein-encoding RNAs. They have transcended time as they continue to replicate within eukaryotic cells, mostly in plants. None of them have shown RNA polymerase activity. The *Pospiviroidae*, represented by the potato spindle tuber viroid (PSTVd), and the *Avsunviroidae*, by the

* When we refer to a viroid as a living entity, we mean the system made up of the RNA strand and the intracellular structures and processes that support it.

avocado sunblotch viroid (ASBVd), form the two main families of viroids. They both use a common rolling-circle mechanism for replication, a strategy supported by their circular structure. Through this mechanism and under the catalytic action of an RNA polymerase enzyme, plus-strand RNA templates first produce minus-strand multimers from which new plus-strand copies are generated.[2] The two families of viroids differ by several characteristics. Members of the *Pospiviroidae* family have a centrally conserved region, display a DNA-like compact folding structure, and replicate in the nucleus using the DNA-dependent RNA polymerase II.[2, 3] They do not have a hammerhead structure for a self-cleavage capacity.

Conversely, the *Avsunviroidae* replicate in the chloroplast compartment of the cytoplasm and display catalytic activity carried out by a hammerhead structure but are devoid of a central conserved region. Another difference between the two families resides in the way they replicate through the rolling circle mechanism. The plus polarity is conventionally allocated to the most abundant RNA strand in vivo and the minus polarity to its complementary strand. The plus-strand RNA multimers of the *Avsunviroidae* are synthesized from previously cleaved and circularized minus-strand monomers. This mode is said to be symmetric because strands of both polarities are replicated in the same manner. In the case of the *Pospiviroidae*, the plus-strand multimers are instead copied directly from the linear minus-strand multimers before strands of both polarities are cleaved into unit-length linear molecules and circularized by host enzymes.[4] This variant of the rolling-circle mechanism is said to be asymmetric.

Finally, the nucleotide sequence determines the primary structure of all viroids. However, while the *Avsunviroidae* display a branched secondary structure, the *Pospiviroidae* adopt a rod-like pattern generated by intramolecular folding that is induced by Watson-Crick base pairing.

In short, the viroids represent the simplest forms of life. We tend to think that at that embryonic stage of life, the first living organisms could not completely take their independence from mineral surfaces for diverse reasons, including the capacity of the latter to concentrate molecules important for life or the polypeptides they produce. Elisa Biondi and colleagues have also demonstrated that clay allows adsorbed hammerhead ribozymes to preserve their catalytic activity.[5]

Satellite RNAs of Plant Viruses

We also identify the satellite RNAs (satRNAs) of plant viruses among the earliest forms of life. Those entities are carried within the capsids of some plant viruses on which they are completely dependent for their replication. They vary in size from approximately 200 to 1500 nucleotides.[6] For their replication and encapsidation, satellite RNAs parasitize plant viruses called helper viruses (HV). Based on their protein-coding capacity, they are broadly classified into two main groups, small and large. The smaller satellites are further classified into linear and circular. They do not appear to code for any proteins. Because of some structural resemblance to viroids, the small circular satellites are sometimes referred to as virusoids. The larger satellites, however, contain open reading frames (ORFs) of which some protein products have been detected in vivo.[6]

Like viroids, satRNAs replicate through the rolling-circle mechanism but with self-cleavage of the multimeric RNA strands, mediated by cis-ribozymes.[7] Indeed, Prody et al. have demonstrated the in vitro formation of biologically active monomeric satellite RNAs from multimeric forms by autolytic processing, i.e., without the intervention of any extraneous enzymes.[7]

The self-cleavage involves a divalent cation, mainly Mg^{2+}, and is a catalyzed transphosphorylation, during which the internucleotide phosphate at the cleavage site is attacked by the 2'-hydroxyl group, yielding on one side a 2', 3'-cyclic phosphate, and a 5'-hydroxyl terminus on the other side.[7, 8] However, in addition to hammerhead ribozymes, some satRNAs display ribozyme activity of the hairpin class responsible for the spontaneous circularization of monomeric RNAs.[2] Though still non-protein-coding, the small RNA satellites seem to represent an evolution from the pospiviroids, for instance. They have ribozyme activity participating in the process of replication.

Both viroids and satRNAs of plant viruses need cellular polymerase enzymes to replicate. We submit that their ancestors replicated under the catalytic action of polymerase enzymes synthesized by mineral surfaces. In the diluted environments of the many prebiotic reactions, the wherewithal for this endeavor was concentrated by substances with a large surface area, such as clay minerals and LDH. However, to take their independence from such concentrators and evolve to a higher organizational level, the first living systems needed to build their own compartments through the synthesis of membranes, a step that would lead to the biological cell.

How then did the first cells form?

We take the view that the first cells were an evolutionary strategy consisting in moving away from the initial rudimentary system of compartmentalization of the chemistry of life. We have shied away from proposed models according to which the initial RNA-protein replicons got encapsulated by spontaneously preformed spherical lipid bilayers. We submit that the first cellular membranes were not natural vesicular liposomes that would subsequently shelter naked organisms, but rather structures built from information encoded in some of the earliest genomes. This hypothesis is also guided by the modus of membrane formation and improvement of modern cells. The making of a membrane was one of the earliest evolutionary steps of the first life forms.

The Making of Membranes

The naked RNA-protein replicons at the origin of modern cells, say we, to express a phenotype above and beyond their basic structure, to evolve from the state of naked organisms to RNA cells, needed to satisfy two major requirements: acquisition of protein-coding capacity and fabrication of a membrane. Overcoming these two challenges will in fact provide them with mechanisms to better leverage the resources available from the primitive hydrosphere, an outcome likely to improve their replicative yield.

Some of the first proteins produced by these replicons under these circumstances were most likely intended for the synthesis of a membrane. We support this assessment with the study of the modus operandi of membrane formation in bacteria, a skill they inherited directly from the RNA cells. What we know is that, according to T. Cavalier-Smith, in bacteria, the genomic DNA and the ribosomes that synthesize membrane proteins are directly attached to the cytoplasmic membrane.[9] This organization implies, in our view, a direct connection of the first cytoplasmic membranes with the genome on the one hand, and on the other with the ribozymes synthesizing their protein components. Later in eukaryotes, this relationship persisted, but in their case, the chromosomes and ribosomes responsible for making membrane proteins got instead attached to the membrane of the rough endoplasmic reticulum. In both cases, the membranes grow by the direct insertion of the individual molecular components, some of the embedded proteins serving as enzymes for the synthesis of the membrane lipids.

We submit that at that particular period of the history of life, and in the context of the amino-acid- and peptide-rich environment of mineral surfaces, the peptide bond formation for living organisms will be transitioned from mineral surfaces to particular ribozymes, ancestors of the ribosomal RNA. This view is largely supported by the concept that ribosomal RNAs are the catalytic components responsible for the peptide bond formation in ribosomes. Those ribozymes will synthesize proteins intended for the construction of a membrane around some of the first organisms. From precursors available in the environment, lipids will be synthesized at some later time by the catalytic action of some of those proteins to give the cell membranes the lipid signature to which we are accustomed. The direct interaction of cellular mRNA with the ribosomes has its origin in the precellular world, where coding RNA sequences interacted directly with ribozymes to build the first membranes, which were proteinaceous in nature.

A given nascent membrane, while attached to the ribozyme, will diffuse around it until a sphere is formed, segregating the system (RNA genome and associated biocatalysts) from the environment. This was the birth of the RNA cell. A membrane enzyme, the Na-K-ATPase, will in time work to differentiate the content of the cytosol or internal milieu of the cell from the exterior. Studies in eukaryotes have shown direct nuclear membrane protein and DNA interactions. One study[10] shows that several transmembrane nuclear envelope proteins bind directly to DNA and suggests the critical nature of those interactions for the initial assembly of the nuclear envelope. This theory of the genesis of the plasma membranes can find, as mentioned above, some supporting evidence in the membrane formation of both prokaryotes and eukaryotes, having themselves inherited their membrane organization, either directly or indirectly, from the RNA cells. We believe that, while phospholipids are considered the primary building blocks of most biomembranes, proteins preceded them in this endeavor. It is therefore preferable to consider biomembranes as protein structures with intercalated lipids.

The very first cellular membranes having been exclusively protein in nature, one would think that they could have been synthesized by mineral surfaces, but instead the evidence shows that they stemmed directly from the evolution of the genomic material.

Evolution Toward Protein-Coding Capacity

Evolution toward protein-coding capacity is seen in the satellite RNAs of helper viruses that infect plants and humans. Their counterparts that became modern cells evolved the genes to code for proteins with hydrophobic components destined to the formation of an exclusively proteinaceous protomembrane. Thus were formed the first protocells that will evolve into the cells we know today.

Large Satellite RNAs of Plant Viruses

The large satellite RNAs differ from the small ones not only by their size but also by the fact that they contain ORFs and encode proteins that may play a role in either their replication or pathogenesis.[6] Rubino et al. have reported a region of homology among the proteins encoded by three large satellite RNAs of the nepoviruses.[11] The specific protein of sTBRV, sGFLV, and sCYMV-L1, respectively the large satellite of the tomato black ring nepovirus (TBRV), the grapevine fanleaf nepovirus (GFLV), and the larger of the two satellites associated with the chicory yellow mottle nepovirus (CYMV), were compared. They all contain a particularly basic domain found in the first one hundred amino acid residues. The basic domains confer to those proteins a histone-like property, and an RNA-binding role has been cited for them.[6, 11] Furthermore, the hydrophilicity profile of the sCYMV-L1 protein shows a hydrophobic N-terminal domain followed by a highly hydrophilic region. The mode of acquisition of the ORFs by the large satellites is not clear and we can only offer some conjectures. One possible pathway could be through the rolling circle replication mechanism of these RNA replicons. This mode of reproduction is a process at the end of which multimers of the initial RNA template are generated. Those multimers are subsequently cleaved into monomers, which are then circularized. This mechanism can potentially generate genomes that are larger and that displays significant genetic variability, the latter feature due to the low replicative fidelity of RNA polymerases. Open reading frames can in this way be added to previously non-protein-coding RNAs.

Hepatitis D Virus (HDV)

The hepatitis delta agent (satellite RNA of the hepatitis B virus [HBV]) displays some of the same properties of the satRNAs of plant viruses. Like them, it carries its own ribozyme domain and requires a helper virus for the provision of an envelope necessary for virus assembly.[12] We will subsequently refer to this agent as hepatitis delta virus (HDV), its most common appellation.

HDV seems to replicate through a symmetrical rolling-circle mechanism during which it reproduces not only a new genome but also a complementary RNA strand known as antigenome.[13] This complementary or antigenomic strand of the hepatitis delta virus RNA contains an open reading frame responsible for the synthesis of a protein called the hepatitis delta antigen (HDAg), also sometimes referred to as delta protein or delta antigen. This protein exists in two forms, labeled small (S-HDAg) and large HDAg (L-HDAg), containing 195 and 214 amino acids, respectively.[12] The large delta antigen is formed by a 19 amino acid extension to the small isoform. The structure common to both isoforms is basically an amphipathic molecule: an essentially hydrophilic protein with a hydrophobic core in its middle. Zuccola and collaborators have expressed that the peptide corresponding to residues 12–60 of the delta protein forms what is called an antiparallel coiled-coil domain where some hydrophobic residues participate in arranging an extensive hydrophobic core.[14] They also suggested that the N-terminus of the delta antigen protein, common to both isoforms, may play a role in binding the viral RNA, acting as a sort of "clamp" around the molecule.

In summary, this common structure has two remarkably interesting properties. First, it is an amphipathic molecule that binds RNA. Secondly, while mostly hydrophilic, it displays a hydrophobic core toward the N terminus, which also binds the RNA, thanks to its highly basic domains.

In addition to this common structure, the large HDAG not only has an additional 19 amino acid extension at the C terminus but is also isoprenylated at this end. Isoprenylation in the context of the large delta antigen protein is a post-translational modification that involves the attachment of a farnesyl group to the sulfhydryl group of a cysteine residue. This modification is considered essential for the assembly of the virus. It appears to play a key role in the interaction of the HDV ribonucleoprotein with the envelope formed by the HBV surface proteins, and it also results in a substantial increase in the hydrophobicity of the protein.[15] The large hepatitis delta antigen is expressed only later in the life cycle of the virus when an RNA editing event allows translation to continue for the additional 19 amino acids. Not only is it important for the packaging of viral RNA and its interaction with the envelope, but it also inhibits HDV replication, while contrariwise, the small version of the protein is essential for HDV replication.

The opposing role of the two forms of the delta protein and the difference in their respective structure seem to suggest that the RNA editing event could have been an evolution of a subviral entity when it had to seek shelter inside the HBV envelope. The large antigen seems to have been expressed when, in the context of its sheltering inside the HBV, replication had to be repressed. Inhibiting replication would not make sense until such a time. Accordingly, the modification got underway for this purpose. It is possible that for much of the life of the ancestor of HDV, it only expressed the small delta protein.

Discussion

The large satellite RNAs of plant viruses evolved from non-protein-coding entities to replicons that encode and express amphipathic proteins. The hepatitis delta virus also encodes amphipathic proteins. In both cases, the proteins are attached to the RNA genome, forming a ribonucleoprotein structure.[10, 16, 17]

We submit that the satellite RNAs of plant viruses and the HDV virion particle, surrounded by amphipathic proteins, represent caricatures of the first protocells of life. Obviously, they did not evolve to fulfill the full requirement of a cell, but we refer to them to show the earliest steps taken by some of the first life forms toward the formation of the cell. The separation of an interior milieu from the exterior by an amphipathic membrane heralded the most fundamental feature of the modern cell. The ancestors of ribosomal RNA, themselves ribozymes, synthesized those amphipathic proteins[18] according to instructions from the genomic ORFs of these entities. Those early ribozymes, we submit, were reinforced by prebiotically synthesized peptides.[19, 20] Furthermore, as it has been demonstrated from the works of many scientists, including Thomas Cech, the ribosomal RNA evolved from an ancestral peptide-synthesizing ribozyme.[21-23] An experiment conducted by Noeller and collaborators supports the notion that the ribosomal RNA may be responsible for the peptidyl transferase activity of the ribosome by showing that ribosomal specimens of selected bacteria retain much of their activity after vigorous protein extraction procedures.[18, 24] Peptide bond formation during cellular protein synthesis occurs at the peptidyl transferase center (PTC) of the large ribosomal subunit. No proteins were detected closer than about eighteen angstroms to the PTC, which is formed exclusively by RNA molecules.[22, 23, 25, 26]

The preceding arguments are in support of the contention that those RNA-encoded proteins were synthesized by ribozymes.[18] As we have seen, all the substrates were initially concentrated in the interlayer regions of minerals that served as compartmentalization structures. The improvement made by some of the first organisms was to encode and synthesize an amphipathic proteinaceous membrane for self-encapsulation. Proteins intended to become membrane proteins will be synthesized one amino acid at a time and will circumscribe the RNA-protein circuit, from their attachments on the RNA genome, thus forming the first RNA protocells. The lipid components of the membranes, synthesized by the enzymatic action of membrane proteins, will come later. With their respective biocatalysts, a satellite RNA of a plant virus within its protein shell and an HDV RNA molecule contained within an envelope made of the small hepatitis delta antigen were some of the first protocells, the RNA cells (Figs. 13a and 13b).

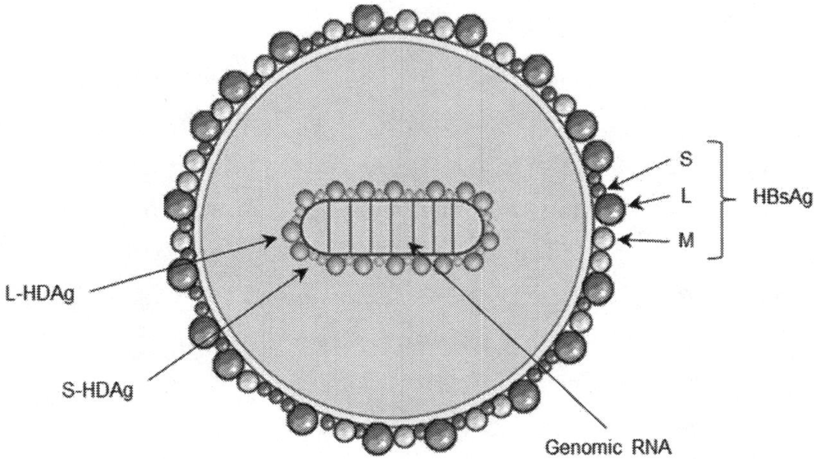

Figure 13a: A diagram of the hepatitis D virus (HDV). The viral envelope contains the three forms (small, medium, and large) of the envelope protein of the hepatitis B virus (HBV). Inside the envelope is the HDV genome: a circular, single-stranded RNA (ssRNA) surrounded by the two forms of delta antigen (L-HDAg and S-HDAg). Reproduced from Turon-Lagot et al.[27] Used under the terms of their Creative Commons Attribution License.

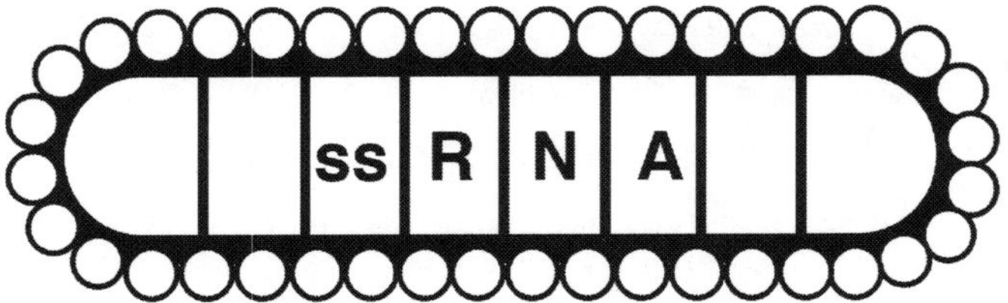

Figure 13b: Partial schematic representation of the ribonucleoprotein complex formed by the proposed ancestor of HDV, which we suggest constitutes a prototype of RNA cells. Here are represented the HDV ssRNA genome and the small delta protein. (NB: partial representation only, as supposed other structures such as RNA polymerase enzymes are not represented.) Adapted from Turon-Lagot et al.[27] Used under the terms of their Creative Commons Attribution License.

Evolution of the First Membranes

The RNA cell so far described is circumscribed by a protein membrane and is centered by an autocatalytic system containing a genome displaying protein-coding and non-protein-coding segments. The exclusively proteinaceous protomembrane, however, will later be replaced by a bilayer structure of mostly lipid components synthesized under the enzymatic action of some of those same membrane proteins. The lipid bilayer will in fact become the signature feature of cellular membranes—proteins mostly relegated to other functions within the membrane, such as providing enzymatic catalysis or acting as communication channels, etc. The lipids of cellular membrane are amphipathic in that they have a polar or hydrophilic head and a nonpolar or hydrophobic tail.[28]

Phospholipids, thus called because of the presence of a phosphate group in their hydrophilic "head," constitute the most common types of lipids in biological membranes. The phospholipid bilayer is a two-layer lipid sheet with hydrophilic heads facing the aqueous environments on either side, while a hydrophobic core is formed by inversely oriented fatty acyl chains. The polar head group contains the backbone, a phosphate group, and a variable group of which choline, ethanolamine, serine, and inositol are some examples. Phospholipids can have a glycerol backbone, in which case they are called phosphoglycerides (Fig. 14), or a sphingosine backbone that, therefore, would classify them as sphingolipids. Phosphoglycerides make up the most abundant lipids in biomembranes.[28-33]

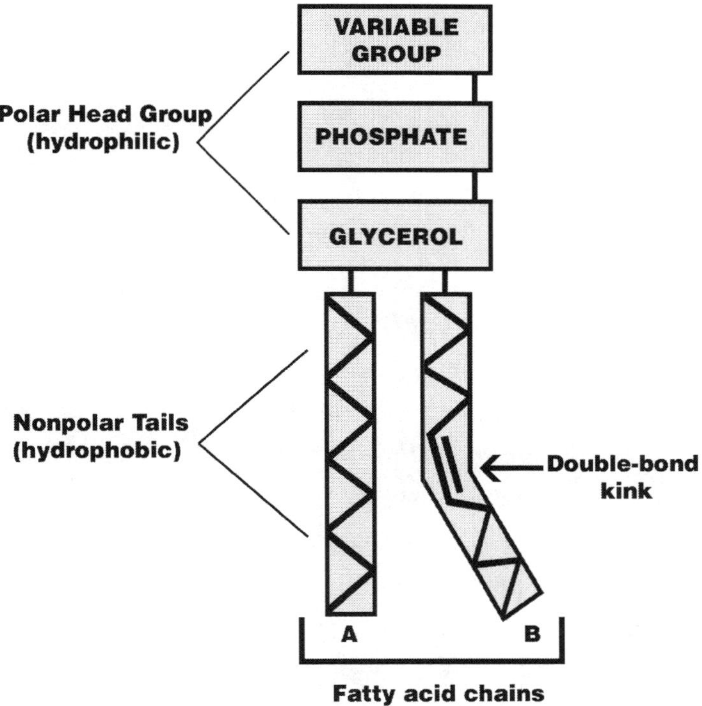

Figure 14: Structure of a phospholipid molecule (A: Saturated fatty acid chain; B: Unsaturated fatty acid chain).

However, the plasma membranes of eukaryotes contain other types of lipids such as glycolipids. Large amounts of cholesterol are also found in animal cell membranes.[28] Furthermore, some bacteria can synthesize non-phospholipid membrane lipids, which by definition lack a phosphate moiety in their structure. Ornithine lipids, sulfolipids, glycolipids, and hopanoids, to name just a few, are examples of such membrane lipids.[30] This is in most cases a method of bacterial adaptation to low phosphorus conditions.[30]

Membrane lipids make up an essential part of all biomembranes. By their amphipathic nature and their better resistance to hydrolysis relative to proteins, they represent the proper building block for the erection of a cellular permeability barrier in the aqueous environment of the earliest periods of life. However, as we have seen, they were not readily available for the initial membranes of the RNA cells. Their *de novo* synthesis required the type of catalysis best performed by proteins. Hence their apparition after a protein membrane was built around the genetic material. Those protein enzymes catalyzed the biosynthesis of the membrane lipids from substrates already present in the aqueous environment and collected in the cytosol. The membrane lipids of modern prokaryotes and eukaryotes are also synthesized from substrates under the catalytic action of membrane proteins wedged among those lipid molecules. They are synthesized at the interface of the cytosol with the cytoplasmic or endoplasmic reticulum membrane in bacteria and eukaryotes, respectively. This is basically the general mode of membrane lipid synthesis. However, advances made since the use of *E. coli* as the benchmark for the study of bacterial membrane lipids have shown that a so-called typical membrane lipid composition does not really exist. Bacterial membrane lipids vary widely both among and within the species, most often depending on environmental constraints.[30]

Nonetheless, with the completion of a membrane around the nucleoprotein system, the RNA cell was born. The synthesis of the cytoplasmic membrane conferred to the cell its essence. The membrane provided the cell with a boundary, differentiating the inside from the outside. Some of the membrane proteins will serve as communication pathways between the cytosolic and exoplasmic faces of the membrane, enabling the movement of substances from one side to the other, thereby regulating the content of the internal milieu. The membrane will take on more tasks, such as photosynthesis or acting as an intermediary in protein synthesis and processing. The cell was then constituted of the cytoplasmic membrane and the interior aqueous milieu that contained the genome, substrates, biocatalysts, etc. The protein-synthetizing ribozyme will in time (likely not before the evolution from RNA to DNA cells) evolve into a mature ribosome that will improve the biosynthetic activity of the cell to an unprecedented level.

Consistent with the error-prone nature of RNA polymerases, the membranes will also diversify. By different organizations of the cytoplasmic membrane and the characteristics of a cell wall, the RNA cells earmarked to evolve into bacteria, for instance, will diversify into groups of which two are the most prominent. One will develop a multilayered murein cell wall, while another will have, sandwiched between an inner and an outer membranes, a thin murein wall. They will constitute respectively the Gram-positive and the Gram-negative cells, precursors of bacteria with similar properties.[30, 34]

CHAPTER

11

Origin of the Cellular Genome

Originally naked organisms whose central structure was an autocatalytic RNA-protein circuit evolved to become enveloped by a protein membrane. Such was the genesis of the first living cells, the RNA cells. The latter, we contend, are the ancestors of all cellular organisms and are involved in more than one way in the genesis of the first viruses. The genome of the RNA cells—and of all organisms, for that matter—evolved from the earliest RNA strands. The genomes of prokaryotes and of eukaryotes are made of genes separated by intergenic noncoding sequences. In eukaryotes, genes are composed of introns and exons which, to simplify, are non-protein coding parts and protein coding parts respectively. Prokaryotes, however, have intronless genes, i.e., exclusively made of exons. The discovery of pre-mRNA alternative splicing, however, has shown that this subdivision of genes may be oversimplified. In alternative splicing, a single gene codes for different proteins by combining the exons in different ways or a unit can be expressed either as an intron or as an exon depending on the cell type. Yet, we contend that these genomic segments, namely the intergenic sequences, introns, and exons, are all derived from the earliest genomes. The RNA cells inherited and further evolved them.

Through comparative genomic analysis, Avina-Padilla and colleagues have identified eighty-one pospoviroid-like conserved sequences of nineteen to twenty-one nucleotides long in the

Arabidopsis thaliana genome.[1] *Arabidopsis thaliana* is a small, widely studied flowering plant that is considered a model organism in plant biology.[2] The hits identified by the team led by Avina-Padilla were located in both coding and noncoding sequences such as introns. The investigators further mentioned that by their comparative bioinformatics analysis, they were also able to identify a tomato gene, labeled *SolWD40-Repeat*, that contains sequence homologies with segments of pospiviroid-derived RNA.[1] One possible interpretation of those findings is that the genome of some eukaryotes seems to have stemmed from sequences of the earliest RNA replicons, such as viroids, and that the sequence homologs include exons and other noncoding sequences, such as introns.

The introns themselves are sequences of eukaryotic genetic materials, generally non-expressed, that separate the expressed codons or exons in a gene or a primary transcript. They are usually excised from the precursor mRNA (pre-mRNA) before the translation of the mRNA strand into proteins. Again, the process of alternative splicing has shown that the difference between introns and exons in terms of their expression into a peptide sequence is only relative. Several studies and insights have also suggested that the introns are related to viroids. Hence the proposition by Diener and others that viroids may be escaped introns.[3-5]

Both introns and viroids are small circular RNAs, remarkably similar in size. For instance, the potato spindle tuber viroid (PSTV) is made of 359 nucleotides, while the intron from *Tetrahymena thermophila* contains 399 bases.[4] The hallmarks of the group I introns are a sequence of 16 nucleotides called group I consensus sequence, and three sets of complementary bases that are also shared by the viroids and virusoids.[5] Some segments of the central conserved region characteristic of the *Pospiviroidae* are also found in group I introns.[4] Group I introns are generally self-splicing ribozymes found in some nuclear, mitochondrial, or chloroplastic RNA genes.[4, 6]

The relationship analysis between viroids and introns continues with the structural genes of eukaryotes that contain the spliceosomal introns. Small nuclear RNAs (snRNAs) in association with proteins form complexes called small nuclear ribonucleoproteins (snRNPs) which are essential components of the spliceosome. In eukaryotes, the spliceosome carries out the splicing of pre-mRNAs, removing the introns and adjoining the exons to form the mature mRNA. The latter is then transported from the nucleus to the cytoplasm to be translated into a polypeptide chain. The 5' end of U1-snRNA, one of those small nuclear RNAs, contains sequences that are complementary to the ends of introns.[3] Diener has demonstrated that the 5' end of U1-snRNA could, by canonical base-pairing, form a quite stable complex with a specific nucleotide sequence of the complement of the PSTV.[3] At the very least, those findings suggest a strong relationship between viroids and spliceosomal introns. However, one cannot either rule out a phylogenetic relationship between those entities, particularly since viroids ordinarily display extensive internal base pairing. In the context of the evolution of genetic information as presented in this project, we believe that introns are evolved forms of viroids, or better, virusoid sequences expressed in cellular genomes.

This evidence of the genetic similarity of cellular organisms with viroids is not exclusive to the components of the genes. Such a comparison has also been made with the stretches of nucleotides intercalated between the genes. The intergenic regions, which, as their name suggests, are sequences of genetic material that separate one gene from the next, make up about 50% of the genome of most animals.[7] The term "intergenic spacer" is particularly used in the context of ribosomal genes. Schmidt-Puchta and colleagues report several regions of "limited similarity" between the PSTV genome and the intergenic spacer (IGS) of the tomato ribosomal DNA.[8] Furthermore, the transposable elements, which, as we have proposed, descended from viroids, populate the intergenic regions of the genome of both prokaryotes and eukaryotes.[9-12] In fact, Silva and collaborators report the presence of conserved fragments of transposable elements in the intergenic regions of both the human and mouse genomes, some displaying high similarity between the two lineages.[12]

Moreover, Michael Kiefer and colleagues have shown evidence of structural similarities among viroids of the *Pospiviroidae* family, retroviruses, and transposable elements.[13] They point out that the end of retroviral proviruses or transposable elements displays striking similarities with viroids of the PSTV group. They also demonstrated that the upper portion of the central conserved region of the PSTV viroids studied displays inverted repeats with dinucleotides U-G and C-A respectively at their 5' and 3' ends. Correspondingly, all retroviral proviruses and certain transposable elements end with the dinucleotides T-G and C-A.[13] To be noted here is that the uracil base in the viroids' RNA genome is replaced by thymine in the DNA of retroviral proviruses. Among other noted structural similarities is a long stretch of purine bases present in the pospiviroids studied, which evokes a similar pattern at the U3 boundary of proviruses. This stretch of bases is the putative reverse transcriptase primer sequence for (+) strand DNA synthesis.[13, 14]

In summary, one can retrace the origin of the different elements of the genome of certain organisms to the viroids and virusoids. Initially non-protein-coding structures evolved to constitute the elements of the genes and the intergenic regions formerly referred to as "junk DNA." And at the root of all this evolution was a simple RNA strand that contained sequences found today in pospiviroids.

We surmise that the pospiviroid genome exemplifies a simple model of the genome of all living systems. Some of the first living entities initially evolved with the addition of protein-coding segments to a genome that did not code for proteins. The supposed pathway of this evolution is through the rolling circle mechanism of replication aided by an error-prone RNA polymerase. This RNA polymerase, which generates multimers from the replicative template, provided only a fraction of the catalytic activities required for the replication of some of the first genomes. Those original RNA polymerases, we contend, were essentially mineral-synthesized polypeptides. The *Pospiviroidae*, but not the *Avsunviroidae*, are devoid of two other catalytic activities needed for replication; namely, a ribonuclease for the cleavage of the multimeric strands into linear monomers, and an RNA ligase to circularize the monomeric units. They co-opt enzymes from their hosts for these important reproductive functions. The ancestors of RNA bacteriophages appear to have been among the first organisms to encode their own RNA polymerase enzyme or replicase.

CHAPTER 12

The First Viruses

Single-Stranded RNA Bacteriophages

The development of the RNA cell provided an opportunity to some naked RNA structures that have not evolved to this level of complexity to use it for their replication. Such was the origin of the first viruses. They had an RNA genome but were not retrophages. RNA bacteriophages therefore evolved from RNA replicons that subsequently obtained their capsids by directly infecting cellular structures, such as RNA cells or bacteria. Viruses are life forms made of an encapsidated genome. They parasitize cellular structures in order to access their stock of nucleotides, amino acids, and proteins, allowing them to reproduce. They come in a variety of sizes, from filterable dimensions to *Megavirales*.

RNA phages like the single-stranded MS2 have the smallest genomes. Like the viroids, their genomes display a high degree of internal base pairing. The MS2 virion, one of the best-studied ssRNA bacteriophages, reproduces within male or pili-expressing *E. coli*. Its genome, to which is attached a copy of the maturation protein, is encapsulated by a naked capsid.[1]

The RNA polymerase of MS2 is made of four proteins, of which only one, the replicase protein, is encoded by the phage genome. The other three proteins are cellular- or host-derived. The virus-encoded RNA replicase is the most essential element of the RNA polymerase activity of infected cells. In fact, once the synthesis of a new RNA strand begins, it can proceed without the help of the three cellular components of the polymerase, since the replicase protein has polymerase activity by itself. In vitro studies have shown that a complex consisting of the phage RNA and the maturation protein is sufficient to infect male *E. coli*. It can then bind to pili in the absence of the capsid and guide the RNA genome into the cytoplasm.[1, 2] This finding supports the idea that these viruses originated from the infection of gram-negative bacteria or their RNA cell precursors by advanced RNA replicons.

Phages with Segmented Double-Stranded RNA Genomes

The phylogeny model of MS2 does not seem different from that of the RNA phages with segmented double-stranded genomes. This family of phages named the *Cystovirida*e is epitomized by the bacteriophage ϕ6, isolated in 1973.[3] Bacteriophages ϕ6 and other members of the *Cystoviridae* group have a genome of three segments of double-stranded RNA, designated L, M, and S. The genome is contained within a core or procapsid made of four proteins labeled P1, P2, P4, and P7. This structure is packaged within a polyhedral nucleocapsid shell made of the protein P8 in all members of the group except ϕ8, where this protein is a component of the membrane.[3, 4] A lipid-containing membrane derived from the host surrounds the nucleocapsid. In vitro studies have shown that procapsids of ϕ6 derived from complementary DNA (cDNA) copies of genomic segment L are competent to recognize, package, and replicate RNA properly with no accessory factors from the host cells.[3-5] Complementary DNA is DNA synthesized by reverse transcription from a single-stranded RNA template, most usually the messenger RNA (mRNA), which is the mature or intron-free transcript.[6, 7]

We interpret this piece of data to mean that at some point in time, a structure made at least of the procapsid and the three-segment genome was an independent, self-replicating living system. The procapsid carries the protein P2, which is in fact the RNA-dependent RNA polymerase responsible for RNA strand synthesis. The procapsid served also to compartmentalize the genome, which is bound to its innermost protein, P7.[8] P4, the most outer protein of the procapsid, contains a hydrophilic C-terminal tail, which is located at its most outer aspect because it interacts with the nucleocapsid.[9] It also contains a hydrophobic core that ranges approximately from aa1 to aa73 and from aa153 to aa247, not accessible at the outer surface of the procapsid. The procapsid, we posit, then formed a water-impermeable membrane for the RNA genome, and the complex was therefore an RNA protocell just like the ancestor of the hepatitis D virus. This structure, we assume, infected a host that was a pseudomonad precursor. The expression of a capsid shell within its host afforded it the status of a virus.

The complete phenotypical description of the ancestor of ϕ6 when it initially infected its host and the details of their interaction at that particular time are unknown. However, nucleocapsids obtained by incubating P8 with procapsids containing inside genomic segments of ϕ6 are capable of infecting host cell spheroplasts.[4] One can always assume that the bacteriophage and its host have coevolved and that, at a point in their past, the RNA segments, packaged to a certain level, were able to be transmitted to a cellular precursor of the bacterial host.

CHAPTER 13

The Split of the RNA Cell

Nothing is born or perishes, but already-existing things combine, then separate anew.

—Anaxagoras

Our proposal for the origin of prokaryotes and their DNA viruses consists of the partition of RNA cells into viruses and cell matrices while the latter will become DNA cells by the action of the reverse transcriptase enzyme. This method of virogenesis did not include a lytic cellular process, as it produced a virus along with a DNA cell. Those viruses were then enveloped and, in all likelihood, were retroviruses or retrophages. But what about the details of that partitioning? Was an escaped virus a genomic duplicate or rather composed of specific segments of the RNA cell's genome?

Many clues can be gleaned from current basic biological data. In favor of a duplication theory is the finding that, as reported by some authors, the DNA composition of the Caudovirales generally reflects that of their host bacteria.[1, 2] We propose, at the very least, that segments of eukaryotic genomes generally not found in bacteria, such as introns and retrotransposons, originated exclusively from the genome of their retroviral ancestors. The viruses that escaped from the RNA cells were then not genomic duplicates, rather they contained sequences that generally left no copies in the remaining cells. Those sequences will be ancestors of eukaryotic introns and retrotransposons, for instance.

Due to circumstances yet to be identified, retroviruses will escape from the RNA cells while the latter will be transformed into DNA cells. The RNA cell genome will be partitioned into two entities, one of which is a retroviral component. This retroviral genome contains non-protein-coding sequences and open reading frames necessary for its replication. The retroviral genome will be cleaved out of the RNA cell genome by the action of catalysts, according to a transesterification mechanism similar to the spliceosomal splicing reaction. The liberated retroviral genome, encapsulated within a newly synthesized capsid, will bud out of the cell, which itself will undergo a process of reverse transcription of the remaining RNA genome into DNA. A ribosome will subsequently mature, transforming the cell into a prokaryote. Such an event in the history of life represented a critical transition from an RNA-based storage of genetic information to a more stable, DNA-based, system. The reverse transcriptase enzyme is at the very center of this transition. The escaped viruses, which were retrophages, will in time return to those DNA cells to reproduce themselves. They will subsequently be transformed into DNA viruses of their respective prokaryotic hosts. The split of the RNA cell represents a variant of asymmetric cell division in which none of the offspring is a perfect copy of the parent cell, the retrophage not being a cell at all. Such was the genesis of the original DNA viruses and their prokaryotic hosts or the event at the very origin of gamete cells in eukaryotes.

The Case of the Eukaryotic RNA Transcript

The eukaryotic RNA transcript or pre-mRNA presents characteristics in some ways similar to the RNA cells. Its exons and introns correspond respectively to the coding and noncoding sequences of the RNA cell genome. In fact, the eukaryotic DNA contains noncoding sequences within and between the genes in the forms of introns and intergenic regions, respectively. Contrarily to the intervening sequences of the genome, the introns constitute part of the genes and are thus present in the primary transcript. Before its transport to the ribosome to be translated into a protein, the primary transcript is processed by the excision of the introns and splicing of the remaining exons, a two-stage transesterification reaction mediated in the spliceosome by the small nuclear ribonucleoproteins or snRNPs. The snRNPs are complexes made of proteins and small nuclear RNAs (snRNAs). The snRNAs themselves are highly conserved RNAs found in the eukaryotic nucleus that, through base-pairing with the pre-mRNA, are involved in the recognition of its exon-intron junctions or splice sites.[3] Instead of the protein component of the snRNP, the snRNAs catalyze the splicing. The phosphodiester bonds of the intron-exon are broken in favor of the formation of new exon-exon bonds, leading to a mature mRNA and a cyclized intron. The excised introns assume a so-called lariat structure as previously alluded to, and their fate is still a matter of research.[4, 5]

It is important to note that this transesterification reaction that is central to the splicing process does not require energy input from an outside source. The energy of the broken intron-exon phosphodiester bond is preserved and used for the formation of the new exon-exon bond. The mature mRNA will then be transported to the cytoplasm, where its information will be translated into a polypeptide chain. In summary, it is a process of partitioning of coding and noncoding genetic segments through the rearrangement of phosphodiester bonds.

The pretranslational processing of pre-mRNA is grossly a mirror image of the partition of the genome of RNA cells in that some noncoding sequences are cleaved out, followed by the splicing of the remaining coding sequences. In fact, bacterial mRNA does not typically have introns. The ancestors of introns will bud out of the RNA cells, probably with other sequences, to form prokaryotic viruses, which were retrophages. Contrary to the introns of the eukaryotic pre-mRNA, the retrophage genome contained open reading frames for the synthesis of proteins necessary for its replication when it reaches the host cell. It is accurate, however, to state that prokaryotes and some of their viruses emerged from the RNA cells in a process somewhat similar to the excision of introns and the formation of a mature mRNA out of a eukaryotic primary RNA transcript.

Retroviruses

The process of retroviral replication hosted by eukaryotic cells follows a set of similar circumstances. After entering the eukaryotic host cell by the fusion mechanism, the retrovirus follows an intracellular phase of its life cycle that involves the reverse transcription of its genomic RNA into DNA, forming an RNA-DNA hybrid. The ribonuclease H portion of the reverse transcriptase enzyme digests the RNA template in the RNA-DNA hybrid, leaving intact the proviral DNA. With the help of the integrase enzyme, the proviral DNA is inserted into the host genome through the formation of new phosphodiester bonds in a concerted DNA-cleavage-ligation reaction that, again, is energy self-sufficient.[6]

The proviral DNA, once integrated, can then be transcribed like any cellular gene into a full-length RNA. This viral genome transcript can be assembled into new virions, the details of which are described in most microbiology textbooks.[6, 7] Through this process, the proviral DNA has become part of the host cell genome and is transmitted vertically with each cellular division. The genetic recombination between retroviruses and eukaryotic cells involves transesterification reactions in the reverse direction of the described partition of the RNA cell genome.

The Case of the Archaea

Archaea, as it can be inferred from the specific characteristics of their membranes (see next chapter), evolved from RNA cells other than those that led to bacteria. Archaea are mostly extremophiles, and their RNA ancestors were either different or must have diverged from those of bacteria before or during the appearance of membranes and cells. Archaeal viruses originated in a manner similar to bacteriophages. However, no archaeal RNA viruses have been discovered so far. We submit that Archaea codiverged from RNA cells with their viruses, or at least some of them, in ways similar to the bacterial model.

As their membranes display critical structural differences with those of bacteria and eukaryotes, they do not infect animals or plants. Their viruses are also, by and large, specific to the Archaea domain.[8] Like bacteria and their bacteriophages, Archaea and their viruses have continued to coevolve after their divergence. There is no evidence, however, that pairs of Archaea and related viruses later combined again to generate specific eukaryotic forms of life, as happened with some bacteria and their phages.

What about the origin of the so-called nucleo-cytoplasmic large DNA viruses (NCLDVs)?

The Specific Case of the NCLDVs

The nucleocytoplasmic large DNA viruses (NCLDVs), a term coined in 2001, comprise an apparently monophyletic group of double-stranded (ds) DNA viruses that infect animals and diverse unicellular eukaryotes.[9, 10] Those viruses possess large genomes and, as their name suggests, replicate either in the nucleus or in the cytoplasm of their hosts. The NCLDVs include the families Poxviridae, Asfarviridae, Iridoviridae, Ascoviridae, and Phycodnaviridae, and groups of giant amoebal viruses such as Mimiviridae, Marseilleviridae, Pandoraviridae, Pithoviridae, etc. Comparative genomic analysis provides a compelling argument that those viruses share a common ancestry as indicated by the conservation of a set of shared genes among them.[9, 10] A great number of those viruses seem to share other biological features, such as replication within cytoplasmic viral factories.[11, 12] Based on their common origins, and in reference to the large size of both the viruses themselves and their genomes, they were proposed in 2013 to be classified under the order "Megavirales."[9] But it was with the discovery of the genus Mimivirus reported in 2003 that the debate on virus origins and definition was reopened.[13-15]

The first member of the genus Mimivirus, a virus named Acanthamoeba polyphaga mimivirus (APMV), was isolated in 1992 from the water of an air-conditioning system in Bradford, England, during the investigation of a pneumonia outbreak.[13] APMV is visible under a light microscope and was considered at first to be a gram-positive bacterium. However, to the dismay of the scientists studying the "bacterium," it contained no ribosomal genes that would have allowed for its speciation, the 16S ribosomal RNA gene being the standard for classification and identification of bacteria. It was instead revealed to be a large virus with an icosahedral capsid when submitted to the electron microscope. In the name APMV are included the facts that it was considered a "microbe-mimicking" virus, and that it replicated within the amoeba Acanthamoeba polyphaga.

The excitement that followed the discovery and identification of APMV invigorated the virology world and opened a new era in virus research. Before long, many relatives of Mimivirus were isolated and new families of giant viruses like Marseilleviridae and Pandoraviridae were discovered and described.[16]

Mimiviruses, like other megaviruses, have been found to contain genes that were until recently believed to be exclusively cellular. They include genes associated with mRNA translation, such as several transfer RNA genes.[16, 17] Therefore, they have significant overlapping features with ribosome-containing cells and consequently are not totally dependent on them to complete their replication, as is typical of classic viruses. They have a wide host range that includes single-celled eukaryotes such as algae and protozoa but also insects, worms, and vertebrate animals.[16] And as stated above, APMV is a human pneumonia agent.

When we initially learned about the NCLDVs, we hypothesized that they originated from remnants of RNA cells from which viral elements have previously escaped, as in prokaryogenesis. In other words, they have failed for an unknown reason, after the escape of the viruses from the RNA cell matrices, to muster a complete evolution into ribosome-expressing organisms. As part of this scheme, we expected them to have their own infecting viruses, which has been proven in the identification of "virophages" such as Sputnik that infect Mimiviruses. Multiple other such replicons have been discovered: the Marseille virus has its own virophage, the Ma-virophages replicate within the giant virus named *Cafeteria roenbergensis* virus (CroV), etc. Another virophage, Sputnik 2, infects the megavirus Lentille virus. Sputnik 2 integrates its genome into its host's in a manner analogous to interactions between bacteriophages or retroviruses and their respective hosts.[16]

We therefore believe that these megaviruses came from RNA cells that were partitioned into viral elements and cellular remnants, the latter having obviously failed to complete their evolution into cells proper or bacteria. They parasitized eukaryotic cells for their replicative needs and, in so doing, became encapsidated. These NCLDV contain genes that are phylogenetically related to every domain of life. Moelling reports similarity of the genomic map of amoeba viruses in the order of 56% with eukaryotes, 29% with bacteria, 1% with archaea, 5% with other viruses, and about 10% with no known organisms.[18] While many scientists interpret those data to mean that the megaviruses are chimeric organisms that have acquired their genome complexity by horizontal gene transfer from gene donors of every domain of life,[16, 18, 19] we read them in a completely different light. We believe and suggest that the presence of a certain genetic homology of megaviruses with bacteria, eukaryotes, and classic viruses stems from their status as RNA cell remnants with failed prokaryotic evolution. The Marseille virus itself contains a chimeric RNA-DNA genome, also indicating an incomplete reverse transcription of the direct ancestor's RNA genome.[18]

One question remains, however: if those megaviruses resulted from the partition of their RNA cell ancestors into them and their associated virophages, how then were they able to develop a capsid, knowing that only viruses possess capsid-encoding genes? One answer, we surmise, is that these capsid-encoding genes could have been obtained before partition through gene duplication, a phenomenon that some authors think is prevalent in those viruses.[16] In a more likely scenario, however, these genes would have been provided by lateral gene transfer to the giant viruses during infection by their virophages. This last scenario finds support in the mobile genetic elements called Polintoviruses. Also called Mavericks or Polintons, they harbor in their genome homologs of the capsid protein genes of virophages and of all the NCLDVs.[16, 20]

All in all, the NCLDVs have challenged the paradigm of viruses as filterable, submicroscopic infectious agents. Viruses are best defined as encapsidated organisms. Classic or filterable viruses usually consist of a nucleic acid genome enclosed in a protein capsid forming a nucleocapsid. In some cases, the nucleocapsid is surrounded by a lipid membrane called an envelope. The membranes of those traditional viruses are host-derived lipid membranes acquired by budding, usually through the plasma membrane or sometimes into cellular organelles such as the endoplasmic reticulum, the Golgi apparatus, or the nucleus.[21] Those viruses do not have internal membranes.

Viruses with Internal Membranes

The megavirales have one more common characteristic: they belong to a group of viruses with internal membranes, i.e., membranes between the genome and the capsid. This feature, necessary for their belonging to this group, stems from the fact that they were cellular subdivisions before becoming viruses and getting encapsidated in the process. In other words, all so-called viruses with internal membranes resulted from the encapsidation of structures that were originally whole or part of an RNA cell. The Tectiviridae, which are not megavirales, also belong to this group. The internal membrane is a lipid bilayer in all the NCLDVs except, at least, the Asfarviridae, in which it is exclusively proteinaceous.

The virions of the NCLDVs are assembled within cytoplasmic inclusions called viral factories. It is within those structures that the internal viral membranes present in all of those viruses are assembled *de novo*, instead of their acquisition by the budding of the virus through preexisting cellular membranes, a key distinction with classic viruses.[12] From the inside out, the structure of those viruses includes in successive order the genome, the internal membrane, the capsid, and an external membrane in the case of the enveloped viruses (Fig. 15).

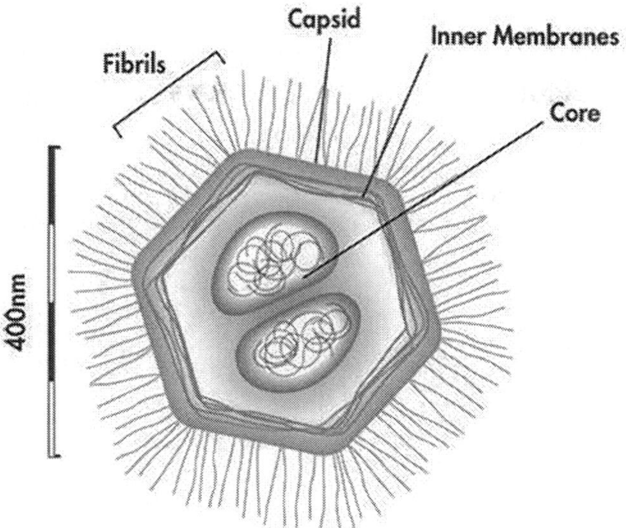

Figure 15: Structure of Mimivirus. Note the position of the internal membranes between the genome and the capsid.

Adapted from Wikimedia Commons and used under the terms of their Creative Commons Attribution License.[22]

We argue that the capsid was a later addition to cellular structures which, for their survival in a changing landscape, had to parasitize more advanced organisms. The internal membranes are chronologically the first ones to have encapsulated the nucleoprotein circuit, thus forming the RNA cells.

The NCLDVs cannot then be classified as either Archaea, Bacteria, Eukarya, or a different domain of cellular life. They gave up the capacity to reproduce autonomously when they adopted the status of viruses. They represent cellular remnants born of the partition of some G3P RNA cells,* after which they became encapsidated. They have undergone reverse transcription but have not completed a full transformation into bona fide ribosome-encoding cells. To survive

* Per opposition to G1P RNA cells (see next chapter).

they had to obtain capsid-encoding genes from their virophages by lateral genetic transfer, so they could access cellular resources to reproduce. However, they had enough genes to survive as autonomous RNA cells in their previous condition.

Finally, while the NCLDVs infect eukaryotes, some DNA viruses with internal membranes, such as those that belong to the Tectiviridae and the Corticoviridae families, are bacteriophages. Their origin is not completely clear. They may have evolved from RNA cells that have not undergone partition but whose genome has been reverse-transcribed.

CHAPTER 14

Membranomics

Comparative genomic analysis has shown that, despite the vast phenotypical differences without and within the eukaryotic lineages, we are more related to each other than our external characteristics disclose. In that vein, it is no question that the great advances made in the study of DNA since the 1950s have fostered a much better understanding of ontogenesis and our common phylogenetic heritage. The domain of phylogenomics, however, as pivotal it has been in affording scientists an aperçu of the genealogy of life, has left many hurdles uncleared. Perhaps the study of membranes, or membranomics, may help to further elucidate some questions on the evolution of the first forms of life left unanswered by comparative genomics or at least consolidate some of its previous phylogenetic decryptions. In other words, phylomembranomics may be able to complement phylogenomics. Indeed, as we have seen in previous chapters, the cellular envelope is the structural component that chronologically succeeded the original nucleoprotein material in the edification of cell-based living systems. It diversified alongside the genome to lead to the earliest subdivisions of life.

Indeed, some characteristics of the envelope structure of cells already constitute a classic method of bacterial classification. For instance, many bacteria are classified depending on whether they take up the stain of Hans Christian Gram (Gram stain) or on the shape of their envelope, such as coccus, bacillus, filament, or rod; these features also provide information on phylogeny. The peptidoglycan component of the internal membrane of Mimivirus is responsible for the retention of the Gram stain.[1, 2] The fact that this giant virus retains the Gram stain is an indication that the formation of the peptidoglycan cell wall goes back to the RNA cells, ancestors of Bacteria, Eukarya, and the Megavirales. Currently, we ignore how widespread this membrane feature is among the Megavirales.

Another membranal factor that can help to categorize life systems is the lipid component of biomembranes, of which phosphoglycerides, we recall, constitute the most common type. The characteristics of membrane phosphoglycerides point to a clear divergence between Archaea on the one hand, and Bacteria, Eukarya, and the NCLDVs on the other (Table 5 and Fig. 16).

Table 5: Characteristics of Membrane Phospholipids according to Life Systems[3-5]

Life systems	Backbone	Layer arrangement	Hydrocarbon side chain	Lipid-glycerol bond	Backbone stereo-configuration
Archaea*	Glycerol 1 phosphate	Monolayer, or "covalently linked bilayer"	Branched isoprenoid chains	Ether-linked	L glycerol
Bacteria Eukarya NCLDV*	Glycerol 3 phosphate	Bilayer	Straight fatty acid chains	Ester-linked	D glycerol

*And their viruses with phosphoglyceride membranes.

As we can see, there are multiple contrasting features between the membranes of those two groups, but the hallmark of what has been dubbed the lipid divide is the glycerophosphate backbone.[3] The backbone of archaeal membrane phosphoglycerides is made of glycerol-1-phosphate (G1P) moieties while glycerol-3-phosphate (G3P) is found in the membranes of Bacteria, Eukarya, and the NCLDVs. Moreover, the respective enzymes involved in the synthesis

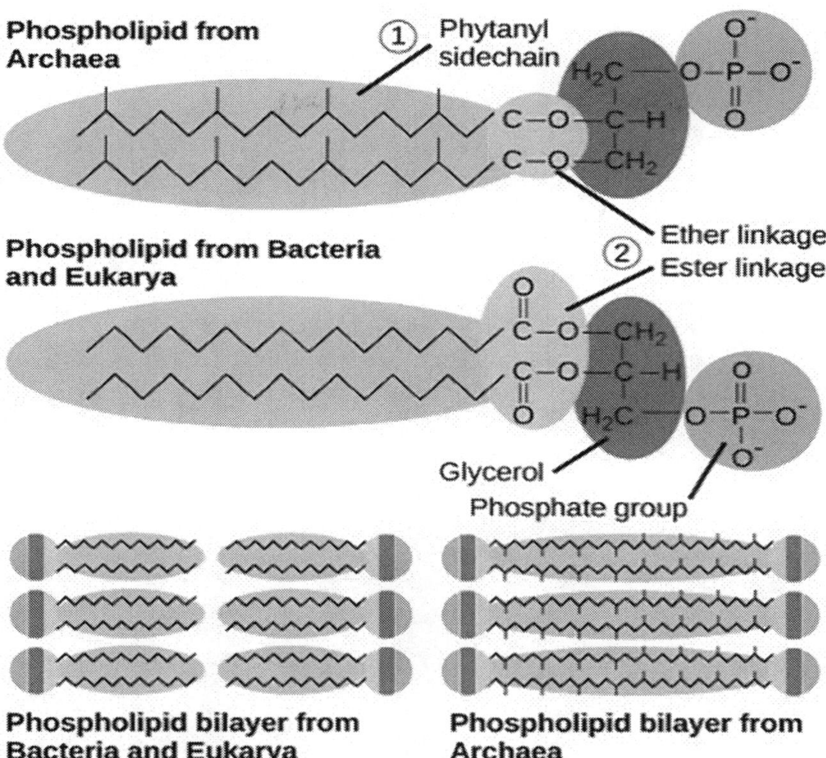

Figure 16: An illustration of some of the major differences between the phospholipids of Archaea on one hand, and those found in Bacteria and Eukarya on the other.[5] Archaeal phospholipids have branched phytanyl side chains instead of linear fatty acid chains. Second, the lipid is connected to the glycerol backbone by an ether bond instead of an ester bond. At the bottom, the covalently linked phospholipid bilayer of Archaea is also shown in comparison to the bilayer found in bacterial and eukaryotic membranes.

of those two specific backbones were found not to be evolutionarily correlated.[6]

Living systems with G1P membranes include G1P RNA cells, Archaea, and their respective membraned viruses, while those with G3P membranes comprise G3P RNA cells, bacteria, eukaryotes, NCLDVs, and their respective membraned viruses. By and large, biologic entities with G1P membranes do not infect those with G3P membranes and vice versa. Archaea may, however, live within the nutrient-rich cavities of eukaryotes.

In Archaea, isoprenoid hydrocarbon chains are linked to the G1P backbone through an ether bond, while in Bacteria, Eukarya, and the NCLDVs, the link between the fatty-acid side chains and the G3P backbone is mediated by an ester bond (Fig. 16). The polar head groups are generally identical for those two major phospholipid arrangements. The structural difference of the membrane phospholipids between those two groups is thought to have resulted from such different environmental conditions of origin, very hostile in the case of archaea, we estimate that if a LUCA ever existed at all as a living system, it was before a lipid-containing membrane was built. We believe that this membrane difference has existed at the very least since the beginning of lipid membranes and has led to the two main types of RNA cells. We will then divide biologic entities into the following categories based on the existence and type of membranes:

- I) Structures without membranes: viroids and non-membraned viruses
- II) Structures with membranes: cell-based organisms and many viruses

There are two main types of membranes:

- A- Embryonic membrane: made exclusively from protein. Ex: hepatitis D virus without the membrane of its helper virus, the hepatitis B virus.
- B- Mature membrane: in addition to proteins, contains lipids that are mostly phosphoglycerides*

* Under growth conditions limited by lack of phosphate, some bacteria build their membranes with phosphorus-free lipids.[7]

The phosphoglyceride membranes are further divided into the following types:

1) G1P membranes
2) G3P membranes

Finally, from comparative membranomics and our theory on the origin of cells and viruses, we have extracted a model of the emergence of the major branches of life which we illustrate in Figure 17.

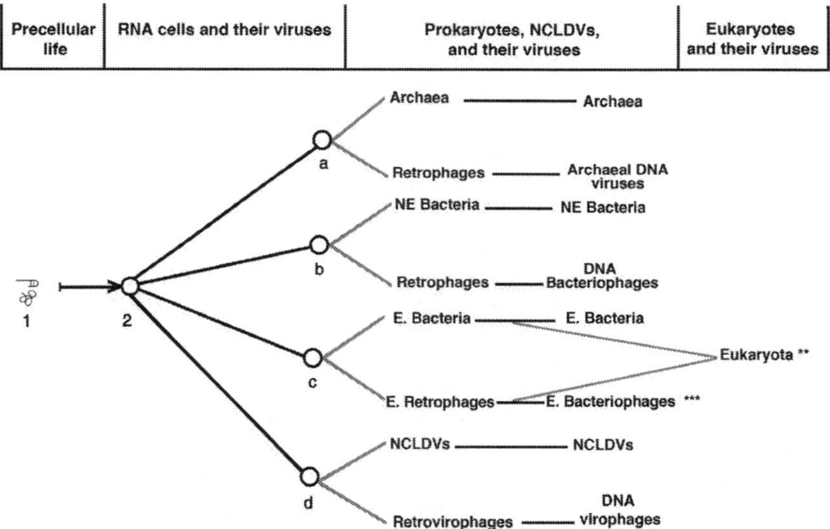

Figure 17: Origin of the major branches of life.

**See Creation of the eukaryotic cell, chapter 1.

***Phages of Eukaryogenic bacteria have yet to be isolated in nature.

1- Ribonucleoprotein system (Viroid and other RNA molecules such as ribozymes + proteins. All from mineral synthesis)
2- Protein-membraned RNA cell

1 and 2 do not necessarily constitute universal common ancestors. They may only represent common features of diverse ancestors. In other words, the spreading of lines from 2 should not necessarily be interpreted as multivergence or arborization from a single ancestor.

Extant homologs of 1 and 2:

- 1: Viroids and small satellite RNAs of plant viruses
- 2: Large satellite RNAs of plant viruses (without their helper viruses), Hepatitis D virus without the membrane of its helper virus, the HBV.

a: G1P RNA cell
b: Bacteriogenic non-eukaryogenic (BNE) RNA cell
c: Eukaryogenic (E) RNA cell
d: Partitionable non-prokaryogenic (PNP) G3P RNA cell
(b, c, and d are G3P RNA cells)

NE: Non-eukaryogenic

Examples of eukaryogenic bacteria: *T. namibiensis*, bacteria of the respective genera *Thioploca* and *Beggiatoa*.

Grey tone lines denote partition or fusion.

- Evolution goes from the left to the right.
- Elements on the left may use contemporaneous systems or those on the right as their reservoirs or hosts where to replicate. For example, satellite RNAs and helper viruses; viruses or bacteria and their eukaryotic reservoirs.

On the Commonality of the Inert and the Living

CHAPTER
15

The Broader Context of Life

Nothing is lost, nothing is created, everything is transformed.*

—Antoine Lavoisier

L'homme n'a point de port, le temps n'a point de rive;
Il coule, et nous passons!**

—Alphonse de Lamartine (Le lac)

The Background

We suspect that life systems and inert matter proceed, each in their own way, toward a similar goal for the universe. We have so far described the phenomenon of life as opposed to inert matter and according to some of its most unique features. However, those defining properties exhibit a purely emergent nature. Life did not appear in isolation, but in the universal context of energy and inert matter. In that vein, we will now dive into the depths of the universe while expressing some thoughts about its origin, makeup, and destiny; for it is our position that the basic features of the universe hold some bearing upon the topic under consideration. We believe that the more we understand about the universe, from energy to

* Paraphrase of the law of conservation of mass.

** Man has no port; time has no shore.
It flows, and we fade away!

inert matter, from black holes to gravity, the more we will be able to discern the purpose of life and its evolution. This chapter, we submit, in addition to making the case for the proposition of the next one, provides meaningful insights into the features of the universe.

The current model of the universe is one of an expanding structure that has emerged from a singularity, through a process dubbed the Big Bang, roughly 13.8 billion years ago. This model has been inferred from the Einstein Field Equations (EFE) for general relativity[1] and later confirmed by, among other findings, the observations of the American astronomer Edwin Hubble. Hubble noticed that the radiations from faraway galaxies in all directions displayed a spectroscopic redshift which, interpreted in the context of the Doppler effect, suggested that they were moving away from us. The discovery of the Doppler effect has taught us that as a light source moves away from an observer, the light waves lengthen and cause a shift toward the red end of the spectrum relative to a reference spectrum of a stationary source.[2] Hubble also observed that the recessional velocity of a galaxy was proportional to its distance from Earth. This observation led to the understanding that the universe was not only expanding but that its expansion was also accelerating.

However, it was George Lemaître, a Belgian priest and professor of physics at the Université Catholique de Louvain who, relying on general relativity and the astronomical data of the time, including some of Hubble's preliminary results, first proposed a theory that the universe was expanding.[3]

Furthermore, Lemaître attributed the cosmological redshift to an expanding space instead of a motion of the galaxies within a fixed space.[4] His paper,[3] published in French in 1927 in the poorly known at the time *Annales de la Société Scientifique de Bruxelles*, remained largely unnoticed. It is said that Einstein, for his part, when faced with the implications of his theories as interpreted by the scientific community, introduced an ad hoc mathematical trick, the cosmological constant (λ), to try to stay consistent with his view of a static universe.[5] Current scientific literature attributes the expansion of intergalactic space to what is called vacuum energy or dark energy. Vacuum energy is likened to a centrifugal force and a "constant density of energy"[6] that is continuously stretching space. The vacuum has a pressure of about 10^{-6} torr, which is about one-billionth of the normal pressure of the earth's atmosphere.[7]

At any rate, according to the current model, the universe evolved to its current state from a singularity through a birthing process called the Big Bang. Two questions are in order. What is the origin of the singularity? And what is the nature of the Big Bang?

A View on the Origin of the Universe

The Origin and Structure of Energy

The universe can be thought of as having originated out of nothingness. It is an idea that, in addition to theologians and philosophers, scientists have grappled with as well.[8] The original idea of creation out of nothing (*creatio ex nihilo*) is found or alluded to in the narrative of the Bible (Macc. 7:28; Rom. 4:17; Heb. 11:3), and is also asserted as doctrine by both Irenaeus of Lyons and Augustine of Hippo.[9] In the cosmogony of Plato, however, the visible world was not created *ex nihilo*. It came into existence when God—whom he calls the demiurge, or the craftsman—brought order to a preexisting chaos.[10] Besides, for several philosophers (Parmenides, Leucippus, Aristotle, Descartes, Spinoza),[11] *nothing* is impossible; therefore, *something* is necessary.

Indeed, the concept of nothing or nothingness is not easy for the mind to grasp. Let us do a thought experiment where we would remove all the matter and radiation (if possible) from the universe. Is what is left *nothing*? Whether we call it the void or empty space or space devoid of real particles, that *something*, which we can describe as indelible or a physical necessity, is obviously not *nothing*.* However,

* One answer to the question of why there is *something* rather than *nothing* is that *something* is necessary.

we will continue to call it the state of nothingness as it is commonly referred to, or simply Nothingness (with a capital N). The state of nothingness from which the universe derives will be here conceived as a spherical structure with a temperature of 0 K. That the universe was pried from a void could not be more counterintuitive. Indeed, it is filled with stars, planets, black holes, and living systems, among other phenotypes of matter or forms of energy. Nothingness* is here conceived as a spherical structure of energy in the form of flatline radiation (Fig. 18a), i.e., in thermodynamic equilibrium. In keeping with the instability of Nothingness, a property of free energy to temporarily self-organize into highly ordered systems or perhaps due to what physicists called quantum fluctuations, the spherical Nothingness suddenly compactified or got condensed centripetally into a "zero size" (Hawking)[12] Big Bang singularity (Fig. 18b). That was the Big Crunch.

Hawking's estimation, however, more accurately represents a universe that is nothing other than the singularity. That Big Bang singularity, which George Lemaître calls the primeval atom, forms the origin (x, y, z) = (0, 0, 0) of a three-dimensional coordinate plane or the center of the current universe in the universe expansion paradigm. As in a quantum superposition model, the immense space of nothingness collapsed or compactified into the smallest possible space, the Planck space or Planck sphere.** The Planck space is a sphere which has neither a center nor a radius but an indivisible diameter of one Planck length. The Planck space singularity is the real Democritean atom of the universe; it is the quantum universe. The Big Bang singularity conceals the totality of energy, time,*** and space. It is the total energy of the universe and, of course, its time, compressed within a Planck sphere. In fact, in our model, those three phenomena constitute one and the same thing. Except down

---

* Nothingness refers to the same thing as the universe in equilibrium.

** Any structure is necessarily three-dimensional and there is no structural measurement that can be less than one Plank length in any direction.

*** The totality of time of the universe is the definition of eternity.

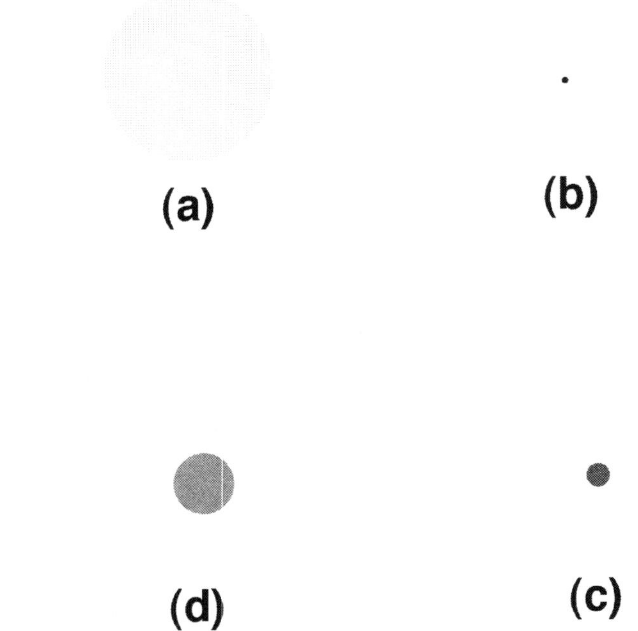

Figure 18: The loop model of the universe.

Four snapshots of the evolution of the universe as captured by an imaginary remote observatory (represented here are only two of its three dimensions): **a**: Nothingness/ Equilibrium; **b**: Centripetal condensation of energy (Big Bang singularity or quantum universe); **c**, **d**: Universe "expanding" with time.

to the last atom, energy is interconvertible with its two fundamental manifestations: time and space. Likewise, the size of space and the frequency of time are closely intertwined; less space corresponds to an increased frequency of time. At the singularity, time and energy are at their highest frequency.

While obviously different, the Big Bang singularity displays some similarities with a Planck star, which, according to Carlo Rovelli, rebounds after being compressed to a certain limit.[13] The rebounding of the universe from a singularity is what has generally been considered to be the Big Bang. This expression carries the implied connotation of a major explosion. However, the transition from Nothingness to an "expanding" universe may have been eerily silent. We can only say that the Big Bang was the beginning of the flattening of electromagnetic (EM) radiation.

The entity that we call the singularity is also the Planck universe. That it was remarkably hot is an established fact. Its temperature should not be far from the Planck temperature of $1.416{,}784 \times 10^{32}$ K, which is theoretically the highest temperature posited by the Standard Model.[14, 15] As assessed by an external ruler, the first Planck time after the Big Bang adds one Planck length of radius to the initial singularity.* A significant portion of the radiation emitted with the Big Bang was high-energy gamma rays, i.e., rays with a wavelength of 10^{-4} nm or probably even shorter. Indeed, according to Wien's law,[16] the wavelength of peak electromagnetic emission of an object is inversely proportional to its absolute temperature: a shorter peak wavelength for a hotter object as compared to a cooler one. Yet the cosmic background radiation (CBR), a remnant of the radiation emitted at the Big Bang, has a temperature of only 2.726 K,[17] and a wavelength of around 2 mm or 2×10^6 nms.** It ensues that the universe cooled down with time and that the wavelength of electromagnetic (EM) waves is not constant but lengthens as they "travel" through the vacuum. This wavelength change may not be apparent even over relatively long distances. For instance, using the age of 13.8 billion years for the universe, the average change from a wavelength of 10^{-4} nm of gamma radiation to the CBR is 1.45 nm $\times$ 10^{-4} per light-year traveled by EM waves. In summary, the wavelength of EM waves emitted by an object lengthens with the distance traveled.

---

* Any reporting of the size or radius of the universe in this work, other than referring to the universe as a singularity, assumes that the measurement is made using an isolated external ruler. The universe proper has neither center nor radius.

** The cosmic background radiation (CBR) was initially detected by Robert Wilson and Arno Penzias in the form of microwave background radiation and measured 3.5 K. The Cosmic Background Explorer (COBE) satellite launched in 1989 by NASA provided a more accurate measure of the background radiation at 2.726 K.[17] The discovery of this radiation is considered another confirmation of the Big Bang transition.

And obviously, their frequency also decreases accordingly. We suggest that those changes in the characteristics of electromagnetic radiations as they travel away from their source be considered the effect of a principle termed the spring law.* The Big Bang transited the initial singularity toward a universe trying to restore the thermodynamic equilibrium previously reversed by the Big Crunch.

Time

What is it that we want to know when we trouble somebody for the time of the day or any similar inquiries? It is simply our position in history compared to a reference point, which itself could be many things including, the previous midnight, the beginning of the year, the birth of Jesus, or even the Big Bang. As for its essence and origin, time is a special property of energy, and the passage of time is nothing but energy in search of equilibrium. Indeed, when set in motion—which is when it becomes evident—it corresponds to the duration it will take a certain amount of energy to reach equilibrium. All subsystems of the universe may have their secondary time, but the time of reference is the one associated with the energy of the big bang singularity. It runs from the Big Bang to thermodynamic equilibrium (Fig. 18b-a). If one can make the argument that time is in constant flux, for practical reasons we submit the Big Bang as Time Zero of the universe. If at this instant—08:58:25 PM EST on 12/31/2020, according to my clock—all the local clocks of the whole universe were to be stopped, the duration since the Big Bang would be the same for each one of them. Time Zero is the same for every galaxy, black hole, or corner of empty space in our universe. Time practically circumscribes the flattening of electromagnetic radiation from the infinite energy frequency of the

* We differentiate this law from the Doppler effect to emphasize its specific stipulation. When one stretches a spring, the distance between the coils increases while the frequency of the coils over a certain amount of space decreases.

singularity to thermodynamic equilibrium or heat death. It is the finite duration (duration here being the key word) it will take to dissipate the energy of the universe and restore thermodynamic equilibrium. It is the succession of separate snapshots that spanned from the Big Bang to the return of maximal entropy. Time is then not an isolated entity. It does not exist by itself. It is an integral part of the different manifestations of electromagnetic radiations. It makes up the internal structure, or rather the spine of EM radiations. Electromagnetic radiations can be thought of as springs in which two consecutive coils are separated by a certain measure of length and a related measure of time. These measures are independent of the distance between the coils as surveyed with an external instrument; they remain constant whether the coils are superposed, stretched to arm's length, or between two towns.

Indeed, the energy of the singularity is stretching out according to a fixed rate, which is considered the speed limit of the universe. This speed limit has been calculated, using rulers and clocks of our own making, to be 299,792,458 meters per second in the vacuum, rounded to simplify calculations to 300,000,000 (3×10^8) m/s. The duration it will take EM radiations from the singularity to achieve equilibrium is then based on that fixed speed.

Space

Space is the necessary socle of energy, or, to use a Platonic word, its "receptacle." It takes the dimension of the distribution of energy. For instance, the Big Bang singularity is the totality of the universe's energy concentrated in the smallest possible space, the Planck space, while thermodynamic equilibrium is the spread of that energy over the maximal reaches of space. Just like light or gravity, space is a manifestation of electromagnetic radiation. Therefore, they all include a time component. Were it not for their inelegance, the appellations of light-time, gravity-time, and space-time would have been more appropriate for those respective phenomena. Our practice of separating them from time has created much confusion. Whenever the concepts of space, light, or gravity are used, their relationship with time should then be implied, and time should not be understood as a standalone phenomenon. Time and space are not two perpendicular coordinates as usually represented in space-time diagrams, but they form a single axis, the spacetime axis. Space or spacetime is made obvious with the stretching of the electromagnetic wave of the singularity. Two of its dimensions result from the perpendicular oscillations of the electric and magnetic fields, while a third dimension is created by the direction of motion of the electromagnetic waves (Fig. 19). That third dimension is a manifestation of time.*

Time represents the frequency of electromagnetic waves whose increasing wavelengths expand space. Time always flows at the same rate. A step of time, no matter the size, is a defined and constant

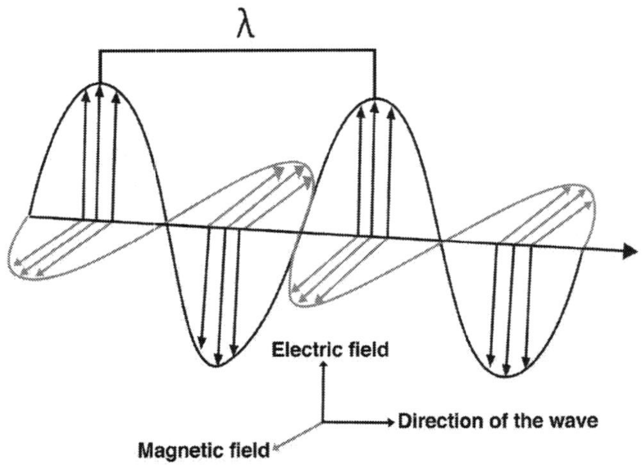

Figure 19: The electromagnetic wave. The electric wave spins vertically while the magnetic wave spins horizontally. Both waves move forward in tandem. The direction and the perpendicular oscillation of the electromagnetic wave create the three dimensions of space. λ denotes the wavelength. The expansion in all directions of EM waves creates a spherical universe.

* EM waves stretch in all directions of the singularity, which then appears to expand spherically.

amount of space. One frequency unit always corresponds to one wavelength unit. In the spring example, for instance, the space between two different coils, whether at rest or stretched out, always corresponds to the same distance. An imaginary observer monitoring from "outside" our universe the rate of expansion of space in any direction of the singularity would find the ratio of 299,792,458 meters per second. This constant speed simplifies the calculation of the radius of the universe at any time after the Big Bang. This radius is the distance from the edge of the singularity to the outer edge of the universe at any point. It is one Planck length at one Planck time of the Big Bang, 299,792,458 meters at one second, and 13.8 billion light-years* when the universe is 13.8 billion years old. The diameter of the universe is a function of time.**

We do not see any objection to the fundamental units of time and length in space being represented, the former by the Planck time (tP), the latter by the Planck length (ℓP), of respective value 5.391247×10^{-44} second, and 1.616255×10^{-35} meter.[18,19] We submit that the fundamental and indivisible unit of space is the Planck space (sP) as previously defined. Like the light quanta of Planck, space and time are of the discrete or granular type. The Planck time is the duration it takes to increase the radius of the universe by one Planck length, or for light to be stretched to such a distance. The Planck time and Planck length are calculated from a direct formula using

---

* One light-year being 9.46×10^{15} meters, the radius R of the 13.8-billion-year-old universe is:

$$R = 13.8 \times 10^{9} \times 9.46 \times 10^{15} \text{ meters}$$
$$R = 130.55 \times 10^{24} \text{ meters.}$$

** What we call the speed of light is rather the mathematical expression of the relationship between time and space. This relationship is made evident by how fast massless particles such as photons and gluons are transported through space. Nothing moves but time. Time is the Atlas of Greek mythology, and matter the flying arrow of Zeno.

the speed of light, the gravitational constant, and what is called the reduced Planck constant.[20] But they can also be deduced from each other, knowing one factor and using the formula c = S/T:

c: the speed of light
S: Space or ℓP
T: Time or tP

Thus, the best definition of the speed of light is one Planck length per Planck time. The speed of light (c) is the constant of proportionality in the relation between space (S) and time (T) according to which (c) is one unit* and is equal to S/T.

Spacetime is a manifestation of electromagnetic radiation. Correspondingly, even though space is continuous, the space around any substantial mass in the universe—whether a planet, star, or black hole—results mostly from the radiation emitted by such an entity. Again, as EM radiations stretch away from a body, their wavelength will increase in a manner inversely proportional to their frequency. Similarly, or consequently, with distance away from a mass, measures of length increase in a manner commensurate to the decrease of the oscillations of time.

An imaginary builder finds himself inside a spaceship in circular orbit O around a black hole (Fig. 20). The spaceship sports a workshop that contains all the paraphernalia needed to build rulers and clocks, including technology that uses a light beam for the standard measurements of duration and length. Using this technology, the imaginary man builds a rigid clock and a one-meter-long steel ruler while in Orbit O. He then moves successively to circular orbits A, B, and C, where he carries out the same jobs as in Orbit O. Upon comparing the rigid rulers, he may or may not be surprised to find that the one-meter rulers get progressively longer from Orbit O to C. Likewise, he will also

* This one unit is also 299,792,458 m/s (rounded to 3×10^8 m/s).

find that he had to build the clocks differently. For instance, instead of 60 oscillations per minute at Orbit O, clock O will need to oscillate more than 60 times at Orbit A for a minute of clock A, and so on.

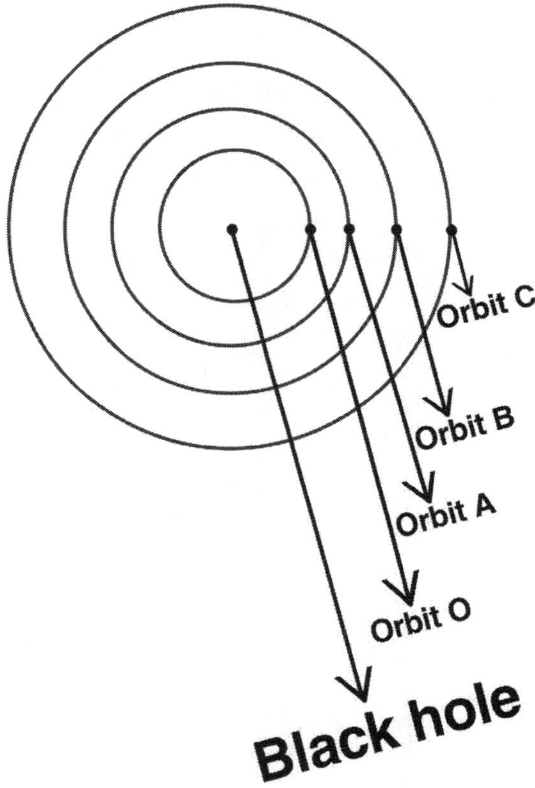

Figure 20: Imaginary circular orbits O, A, B, and C around a black hole.

Clock O built at Orbit O from a black hole and transported to Orbit X further away ticks faster than the local clock.* It is that time, individually extracted from its spatiotemporal reality, seems to slow down as it moves away from a radiation emitter to allow light to stretch

* For instance, the clock O transported to Orbit A would show that one minute and one second have passed, while clock A would show that exactly one minute did. Likewise, were the spaceship occupant to compare the behavior of Clock O at the different orbits, he would find out that it runs slower the closer to the black hole when compared to a fixed duration, say one minute, of the local clocks.

on a seemingly greater distance during the same duration in comparison to a position closer to the black hole. Such is the case that the product of length in space with the frequency of time is always constant and gives the speed of light. Time and space give the appearance of being relative when taken independently. However, they are not separate phenomena. Spacetime is absolute and gives the invariant speed of EM radiation. The relativity of time and space is only an illusion.

Consequently, our imaginary builder will find that the distance/duration relationship, as given by his light beam and in concordance with the relationship between wavelength and the frequency of EM radiation, is constant from O to C. The speed of EM radiation given by the formula $c = \lambda\nu$ (c: speed, λ: wavelength, ν: frequency) is both the universal clock and ruler.

Space and time taken individually are equally stretched in proportion to the distance away from a black hole. Indeed, the galaxies that we can see now from our position will never be out of our reach because of that spacetime/reach of light relationship. Quantized spacetime units can be stretched ad infinitum but their relationship with the reach of light will remain the same.

One more example to illustrate the character of this relationship. While a 1-minute duration is the same everywhere (using local clocks), a given clock G built on the ground floor of a tower will need more than 60 oscillations to complete a minute on the 100th floor and even more oscillations for 1 minute on the 200th floor. Inversely, Clock J built for the 200th floor will reach 1 minute of the 100th floor with fewer oscillations, and with even fewer for 1 minute of the ground floor. Time takes smaller steps (higher frequency, shorter wavelength of EM waves) nearer a radiation emitter than further away.

Rigid clocks destined for significantly different levels of gravity should be built with distinct cogwheel mechanisms for the reasons we have specified. However, *mutatis mutandis*, a process that needs 1 minute for completion on the ground floor of the tower, will take exactly 1 minute anywhere in the universe, using local clocks. We do not age faster at higher altitudes. If rigid clocks are local, spacetime is absolute and universal. This forms the basis of the theory of the invariance or universality of spacetime.

Einstein vs. Hubble

We are then led to one of the major interrogations of the Einstein era, whether the universe is static or expanding. Modern scientists affirm, from their interpretations of available cosmological data, that the universe is expanding and that its expansion is accelerating. We submit instead that the answer to this question depends on the type of ruler used to carry out the measurements. The universe is static and is always a singularity if one measures it with its time-linked, internal ruler, but expanding if measured with an isolated, external ruler. The acceleration of the expansion is only an illusion. In reality, the universe we observe is nothing but the apparent unraveling of the Big Bang singularity. The totality of reality from Big Bang to equilibrium happens within the confines of one Planck space and lasts one Planck time. Another thought experiment might help us illustrate our view on this question.

For your birthday, you are gifted with a robot, one that jumps to move from one place to the next. The automaton is programmed to make 10 jumps covering 10 units of space during 10 units of time. It comes with an N-meter long carpet made of 10 individual units. You gather that each jump will last 1 unit of time and will take our protagonist from one carpet unit to the next.

The carpet units unfold only progressively, one unit at a time, as our protagonist takes its jumps. There is no starting position or "on your marks" zone. The robot is always on a carpet unit. It will be on the first carpet unit when you turn on the game. What you probably ignored but will find out is that, regardless of the extent of deployment of the carpet and of the 10 units of time, the robot still behaves as if each newly added carpet unit completes the whole course while still taking one unit of time to cross it. However, it is the number of carpet units it has already crossed, out of the 10, that informs it of the number of jumps left to do. It can compress its 10 jumps on just one carpet unit during a single unit of time. You turn on the gadget (the Big Bang) and you watch our protagonist take its first jump on the first carpet unit. During one unit of time, it needs to complete 10 jumps on that first carpet unit. It jumps up ten times from that first carpet unit but can only land on it nine of the times as it initially benefited from not having to jump from a starting position. The tenth jump will take it to the second carpet unit, which is added right away. Our protagonist will jump 9 times from the second carpet unit, then 8 times from the third, and so on. When it jumps the last time from the 9th carpet unit, it will land on the 10th unit, which, for now, we will consider as the end of its ordeal.

For you the spectator, the length of the carpet, which is reflected by the position of the robot relative to the first carpet unit, when taken in isolation, progressively increases as time passes in your own rigid clock. Likewise, the radius of the universe, taken in isolation, is increasing with time as measured by our clocks. This perspective reflects the universe expansion model. You also notice that the farther the robot is, the fewer jumps it makes before a new carpet unit is added. The frequency of light decreases and its wavelength increases with time or distance traveled according to the spring law. Likewise, the further away a galaxy is, the faster space seems to be stretching between it and Earth. These observations explain the model of the accelerating expansion of the universe.* These two models constitute the interpretation by Lemaître of the Hubble results.

From the jumper's perspective, however, it has always been on that N-meter carpet, whether it was one, two, or ten units long. It took it exactly ten units of time to cross the ten-unit carpet in its entirety. The carpet never changes size when one uses the internal ruler of the robot-carpet

* The parallel of the robot-carpet apparatus with the universe offers some similarities but is obviously imperfect.

system as a reference. This is the view of the static universe. The universe has a finite and constant amount of energy. The static universe of Einstein was supposed to mean the invariance of its mass and of its energy content, which is an accurate proposition. The unchanging size of the universe is a corollary of the first law of thermodynamics, which states that "the total energy of an isolated system is constant."[21] It can also be deduced from the well-known principle of the equivalence of time and energy.

The expansion of the universe is therefore the result of the passage of time. The apparent acceleration of its expansion follows the same logic of the differential behavior of isolated clocks as a function of their position relative to a radiation emitter. In other words, the expansion of the universe is driven by the motion of time, and its apparent acceleration merely results from the gradual increase in the wavelengths of EM radiations with time or with distance traveled. The mysterious dark energy that pervades the universe has a name: it is the motion of time.

Formation of Matter

Due to the stately unfolding of time, or the fixed speed of deployment of EM waves, the massive energy load of the Big Bang singularity could not at once completely degenerate into flatline radiation. Organized energy remains available for all possible uses. Under the principle of energy convertibility, different phenotypes of matter will materialize, progressing from the elementary particles to all the structures of the cosmos, including living organisms. Order is created locally, which only apparently runs counter to the second law of thermodynamics, but all the while the process increases the entropy of the universe. This was, however, possible only because of the availability of an extraordinary amount of free energy, or energy available to nature to do work. In the context of the infant universe, this energy could not instantaneously and completely dissipate in the form of very disordered heat radiation due to the set speed of EM radiations.

This free energy will interact with empty space to form the atoms of Democritus which include the elementary particles of matter and the force-carrying particles. While Democritus is generally credited with having established the concept of fundamental particles of matter, our account of this idea follows a certain history, one that began in Greece in the 7th century BC with Thales of Miletus who proposed that a single fundamental substance or element was at the origin of all matter. This type of doctrine is generally referred to as monism. Four monistic philosophers of the time will each propose a different element—water, air, fire, or earth—as the single principle at the origin of all things, but Empedocles taught rather that all four of them constitute the ultimate elements of nature.[22] The latter view was subsequently promoted and expanded upon by Aristotle, who added to the scheme a fifth element, the aether.[23, 24]

After the burst of ideas of the monistic and the atomist Greek philosophers, the concept of fundamental particles laid dormant until the early 19th century, when the British chemist John Dalton deduced the existence of the chemical atom out of his Law of Multiple Proportions. Dalton formulated this law from the results of his own experiments and by expanding upon Proust's Law of Definite Proportions. Proust's law states that chemical compounds consist of the combination of chemical elements in defined ratios. But before the end of the century, with the discovery of the electron in 1897 by the British physicist J. J. Thompson, the Daltonian atoms would be demoted from the rank of fundamental particles of matter. Not long afterward, Ernest Rutherford would also discover the nucleus at the center of the chemical atom. By 1932, the basic picture of the atom would be complete with the discovery of the components of the nucleus: the proton, indirectly again by Rutherford, and the neutron, by his student James Chadwick. Currently, the fermions and bosons described by the Standard Model of particle physics are the recognized elementary particles of matter. However, according to string theory, a relatively recent theoretical framework developed to try to resolve the conflict between general relativity and quantum mechanics, those point-like particles of the Standard Model are actually made of a tiny one-dimensional loop named "string." We will leave the string model at that for the moment and continue to consider as fundamental only those particles described as such by particle physics.

As it relates to ordinary matter, two elementary particles belonging to the group of fermions are of utmost importance: the quarks and the electrons. They are the fundamental granular constituents of matter. They give matter its size and shape. Quarks held together by force-carrying particles named gluons will form the protons and neutrons found in the nucleus of the chemical or Daltonian atom. The electrons themselves, according to quantum mechanics modeling, persist around the nucleus in clouds of probability known as "wave functions."[25] A phase of nucleosynthesis sometime after the Big Bang led to the formation of the first few atoms when the newly formed nuclei attracted and retained electrons in their orbits. The atom is the first order of supranucleic matter.

While the elementary particles can be considered formed by the interaction of free energy and empty space,* all supranucleic matter contains empty space. The hydrogen atom, the simplest of the periodic table, will form when a neutronless, single-proton nucleus captures and retains an electron. Ernest Rutherford, as previously mentioned, has greatly contributed to the understanding of the structure of the chemical atom by his successive discoveries of the nucleus and the proton. At the same time, his experiment of shooting alpha (positive) particles toward a piece of very thin gold foil revealed that the atom was made mostly of empty space. More than 99.99% of an atom is nothing but empty space.**[26] Leon Lederman, in his book *The God Particle*, pictures the nucleus at the center of the atom like a green pea at the center of a stadium with a 300-foot radius.[27] Empty space is at the core of the creation of matter.

Matter is then organized into four main levels: infranucleic matter, the nucleus, paranucleic matter such as the electrons, and supranucleic

* Particles obtain their masses by their interaction with the Higgs field of empty space.

** Even though we know that space is radiation.

matter. Infranucleic matter comprises the quarks and their composites.* The nucleus is made of different combinations of protons and neutrons except for the hydrogen nucleus, which has no neutrons. Supranucleic matter comprises all that is made of one or more chemical atoms. Empty space, as already pointed out, is as much a fundamental component of matter as the elementary particles. We submit that empty space is critical in the transition from energy to matter.

If the Big Crunch provides a second layer of an answer to the old question of why there is something more than plain Nothingness,** this answer is further materialized in the post-Big Bang period with the collusion of free energy with empty space to form matter.

Quarks, electrons, and empty space are not enough to form the chemical atoms. Force-carrying particles called bosons are equally important because they mediate the interactions between granular constituents of matter called fermions. Nature is considered to display four fundamental forces or interactions of which three—the strong nuclear force, electromagnetism, and gravity—participate in the organization of matter. The weak nuclear force is responsible for radioactive decay. Like electromagnetism and the strong force, it is described by the Standard Model of particle physics. The Higgs boson, a fifth force theorized in the 1960s, but whose discovery was announced on July 4, 2012, also plays a fundamental role in the organization of matter. It is considered a vibration of the Higgs field through which particles obtain their mass.[28]

The strong force is a nuclear force that holds quarks together to make protons or neutrons and holds together those hadrons within the nucleus of an atom. Electromagnetism involves interactions between charged particles such as electrons and protons. The electromagnetic force holds

---

* The composites of the quarks, which include the protons and the neutrons, are also called hadrons.

** We have already established that Nothingness is describable, that it is *something* and not *nothing*. Therefore, the question of why there is something rather than nothing is a tautology.

the electrons in orbit around the nucleus. Gravity, the weakest of the fundamental interactions, was originally described by Newton, but its currently accepted account is that provided by Einstein's general theory of relativity. Gravity is at work to form galaxies and solar systems.

Gravity

Gravity, while weak, is an interaction that acts over large distances. It results in part from the property of EM radiations emitted by a body to progressively decrease in frequency or lose energy as they are propagated away, according to the spring law. It ensues that the waves and thus space are more folded, the closer they are to a body. The "warping" and "curving" of space in the vicinity of a body are not mechanical actions resulting from the mass of the said body. Those phenomena rather follow the spring law behavior of EM radiations whereby the wavelength of the latter is shortest as they are emitted, but progressively increases as they propagate away from the emitting body. Since the wavelength of a radiation field is inversely proportional to its intensity, this principle is evidenced in the inverse-square law behavior of EM radiations whereby their intensity decreases with distance in a nonlinear fashion and is expressed mathematically by the formula:

$$I_r = \frac{I_1}{r^2}$$

where I_r is the intensity of radiation at r unit of distance from the center of the emitter and I_1, the intensity of radiation at 1 unit of distance. Due to that phenomenon, space is rather folded, if you will, in the vicinity of a body, but progressively thins out with distance.

Objects or radiations interacting with the folded spacetime around a body with a stronger magnetic field must cope with it. A beam of light traveling from a star and passing by another one will seem to deviate toward this new star due to this phenomenon. That beam of light has adopted a new clock about the unrelated star. The spin of a body, combined with the spring law, can make fall on its ground or orbit around it, an object which is not necessarily on its direct path. This is our view of gravity. As the massive body is spinning, the object's position with respect to it depends on the frequency of the successive radiations with which it is met. Progressively increasing radiation frequencies or higher and higher frequencies will bring (accelerate) the object toward the ground of the body. That is why two objects of different masses at the same initial velocity falling from the same position in space, i.e. on the same EM frequency, will hit the ground at the same time. Continuously equal frequencies create a circular orbit. An elliptical orbit is when lower and higher frequencies alternate before the object can hit the ground. Gravity breaks down at the pre-Big-Bang singularity as, by definition, there was no EM emission yet and everything was at the same place. For instance, the inverse square portion in Newton's gravity formula*

---

* $F = G\frac{Mm}{r^2}$

For example, in the case of an object in the gravitational field of the earth
F = magnitude of the gravitational force G = gravitational constant
M = mass of the Earth m = mass of the object
r = distance between the center of the Earth and that of the object.
As we have seen, the universe never ceases to be a singularity. The appearance of things as being separate from each other is nothing but an illusion. The structures of the universe are interconnected by EM radiation as evidenced by the phenomenon of gravity. The square part of Newton's formula for this interaction of nature is the electromagnetic field connecting two interacting bodies. Given an object B separated from another object A by a distance r which is also the distance traveled by the EM waves from one object to reach the other. The orthogonal arrangement of the components of the electromagnetic waves through which the two objects interact forms a square between them. The force of gravity between the two bodies is all the greater as this square is smaller. Hence Einstein's equivalence principle relating gravitation to acceleration.

would yield an infinite value, the distance between the components of a singularity being zero.* Gravity is not a fundamental feature of the universe, but it emerges—albeit quite early in its history—as the singularity decompactifies and creates the structures that we identify as matter. Quantum gravity is an oxymoron.

Fate of the Universe

We contend that the property of energy to temporarily self-organize into highly ordered systems generated the Big Crunch. This same property, as we have seen, is also responsible for the formation of the diverse phenotypes of matter since the Big Bang. However, the second law of thermodynamics is not violated. Indeed, the universe always points to a greater entropy until the reach of thermodynamic equilibrium. Matter, which can be organized at different levels, represents a transitional state between free and disordered energy. A great amount of the original energy is conserved in black holes and stars. The energy of our sun spearheaded the advent of life on Earth. Yet, the destiny of all matter is to render this energy back to the universe in a more disordered form, thereby increasing its entropy.

We must start at the very beginning to grasp the direction of the universe's trajectory. Highly condensed energy that dwindles in an ever-expanding space until it reaches thermodynamic equilibrium at zero degrees on the Kelvin scale; such is the course of the universe. One could stop here and venture to say that's all there is to our universe. One can stop here and venture that this is all there is about our universe. From the Big Bang onward, to the stars and to life, everything will help to reach that destiny. This is the principle that underlies the universality of EM radiation from all matter. Electromagnetic

* The pre-Big-Bang singularity is in fact radiation of zero wavelength and infinite frequency.

radiation is a universal property of all objects with a temperature superior to 0 K. The amount of energy (W) radiated per unit of time from a unit area of a perfect emitter (black body) is proportional to the object temperature (T) according to the formula $W = \sigma T^4$, (σ being the Stefan-Boltzmann constant and T the temperature in Kelvin).[29,30] Stephen Hawking demonstrated that black holes do not violate the second law of thermodynamics in that they emit radiation and thereby increase the entropy of the universe as well.

Living systems radiate energy by the simple fact of being matter and energy. They all radiate as well by their basic metabolism. They radiate even more through activities such as playing, fighting, thinking, getting food, escaping from danger, etc. We humans augment natural radiation through the energy processing machines we create. The common denominator of all matter is thus the inexorable increase in the entropy of the universe through radiation. Up until the heat death of the universe, a law of universal radiation governs all that has derived from the free energy made available by the collapse of Nothingness.

With the emission of electromagnetic waves, all objects lose energy and by the same token lose mass, energy having a direct relationship with mass as shown by the famous Einstein equation $E = mc^2$. Hawking has also demonstrated that the black holes will finally evaporate from their continued radiation emission. Such is the fate of all matter and energy. The universe will undergo what is called a heat death where it will reach maximum entropy or full equilibrium. The smallest sample taken randomly will be representative of the structure of the whole universe. Its temperature will be 0 K. It is a place where nothing will happen. With a frequency of zero quantity, all its energy will have dissipated.* There will be no life, stars, planets, atoms, or particles. The EM radiation waves will flatline.

---

* According to the formula $E = hv$ (E: energy; h: wavelength; v: frequency).

Nothingness will have reconstituted, just to collapse again into a singularity and rebound anew. In the jumping automaton thought experiment, we recall that because our protagonist had benefited from not having to jump from a starting position, it always owed one more jump after completing its landings on a carpet unit. When it lands on the 10th carpet unit from the 9th, it will perform a final jump, while the entire carpet will collapse to a single unit at once, the exercise having been completed. The robot will then land on this carpet unit for a new round of performance. Likewise, the system that begets the universe is cyclic and without end. This is the essence of the loop model that we have depicted in Figure 18. Notice that time is also conserved; the ten units of time have taken the automaton from point A back to point A. We are left to recognize that reason and science have fought and continue to fight a long and arduous battle to expose the facts of our universe. Let us, at this very moment, take stock of their accomplishments. Up this far frontier, they have planted their flag.

The fate of the universe corresponds perfectly, in Greek mythology, to the punishment inflicted on Sisyphus while in hell. The mythological character was condemned to roll a huge boulder to the top of a hill and topple it over onto the other side. However, once near the summit, the unforgiving stone would, every time, roll right back down to the foot of the mountain. Sisyphus is therefore forced to wrestle with the rock again and again, essentially for eternity. Ultimately, his punishment turns out to be a toil that rightly falls into the realm of the absurd, the meaningless, or what Ecclesiastes calls a "chasing after the wind."

Equating the fate of Sisyphus with that of man, Albert Camus, a French writer and essayist of the mid-20th century, in his essay *"The Myth of Sisyphus"*, exhorts humanity to find meaning and happiness in the daily struggles of life. As he memorably put it, "The struggle itself toward the heights is enough to fill a man's heart." Sisyphus, according to Camus, on the way back to his rock, discovers happiness and manages to find an acceptable meaning to his hardship and misfortunes. The message addressed to the living is that the meaning of one's life is what one creates for it. Our sabbaths, however we practice them, are our own trips back to our stone.

The universe is about 13.8 billion years old; it would be interesting to have an idea of the time remaining before its heat death. For this exercise, we will, once again, return to our jumping robot apparatus. We remember that its performance was to last 10 units of time. This duration is also the number of jumps (10) that the robot had to perform on the first carpet unit during the very first unit of time. This was the greatest number of jumps the robot had to perform on a given carpet unit. The universe will then last the number of seconds corresponding to the highest frequency in hertz (Hz) of electromagnetic waves. The highest frequency in cycles per second (Hz) of the waves emitted by the nascent universe is the number of seconds it will last from Big Bang to thermodynamic equilibrium, provided that this principle is not affected by the formation of matter and its mode of radiation. We assume that this frequency is that of Planck, which is also the number of Planck times in one second.* Planck frequency is considered as the highest possible frequency of EM waves. The life expectancy (E) of one cycle of the universe is then 1.855×10^{43} seconds, or about 588 decillion Julian years, according to the following equation:

$$E = \frac{V_{bb}}{Y_s}$$

* An EM wave occupies a dimension of one Planck length in a particular axis. We think the Planck frequency (1.855×10^{43} Hz) is more accurate for this purpose than the highest known frequency of gamma waves (about 10^{25} Hz).

V_{bb} represents the highest frequency of radiation (in Hz) of the emerging universe (at the Big Bang), or the Planck frequency, and Y_s represents the number of seconds in a Julian year.*

As our exercise shows, we have still a very long journey ahead.

The radius (R) of the universe in meters at its heat death will then be the product of the speed of light (c) and time since the Big Bang (T):

$$R = c \times T$$
$$R = 299{,}792{,}458 \text{ m/s} \times 1.855 \times 10^{43} \text{ s}$$
$$R = 5.56 \times 10^{51} \text{ (5.56 sexdecillion) meters.}$$

---

* A Julian or solar year contains 365.25 days.

$$E = \frac{1.855 \times 10^{43}}{31.5576 \times 10^{6}}$$

$E = 587.814028950237 \times 10^{33}$ years from the Big Bang
$E \approx 588 \times 10^{33}$ (588 decillion) years.

CHAPTER 16

The Theory of Evolution by Thermodynamic Guidance

The energy of the Big Bang singularity, to radiate back to Nothingness according to the second law of thermodynamics, needs to create a vast amount of space, the radius of which at any time is itself controlled by the speed of electromagnetic radiations. In other words, reaching thermodynamic equilibrium is limited by the rate at which space is created. Free or highly ordered energy remained within the core of the early universe, ready to participate in the search for equilibrium. One more way besides direct radiation to increase the entropy of the universe is indirectly by the creation of matter. The different forms of matter are, in fact, energy-processing structures (EPS). Their purpose is to store highly ordered energy and progressively process it into less usable or longer-wave forms. Living matter delays decay by stocking and periodically absorbing energy. Within the framework of evolution, the replacement of one structural configuration of matter by another is only permitted if the latter is more efficient in creating entropy. The above encapsulates the mechanics of the universe.

Such is the case, because the purpose of all that will derive from the energy of the Big Bang singularity, from the most elementary particles to the stars and galaxies, from the viroids to man, is to help the universe achieve equilibrium. Life itself is a second-order creation of the Big Bang, arising from interactions powered by ultraviolet radiation and involving, among other protagonists, some earth chemicals such as nucleic acid bases, ribose sugar, phosphate donors, and amino acids.[1]

The free energy of the Big Bang singularity created the universe in which we find ourselves, and we wonder how it all happened. The process is, in some ways, similar to the formation over millions of years of the majestic Grand Canyon by the Colorado River as it flows over a given ground.* Through the floor of mountains and valleys, the kinetic energy of water flowing from a high altitude carved out a ramp to take it from that elevated position to equilibrium on the Gulf of California. The Grand Canyon of today, by size and other features, is more efficient at moving a certain amount of water from high to low elevation than the Grand Canyon of, say, 4 million years ago. Free energy processing is the task at hand. Despite its majesty, the ramp, namely the Grand Canyon, may just be a collateral effect. But I admit that others may have a different interpretation.

At any rate, the processes of the universe are fundamentally teleological, having to do with the imperative to dissipate its energy into disordered forms. We believe and assert that it is also the case of biological evolution. The second law of thermodynamics, or the imperative to increase the entropy of the universe according to the arrow of time, constitutes the driving force of the evolution of the various biological lineages as of that of the whole universe. Briefly, the cardinal law of the universe also governs the emergence and evolution of life. The evolution of a lineage is associated with a better efficiency at increasing entropy. The theory of thermodynamic or entropic evolution states that biological evolution depends on an enzymatic mechanism but is thermodynamically guided. The retrotransposons that mediate the evolution of eukaryotes proceed

* For accuracy, wind has also been a factor in carving the canyon.

through the action of the reverse transcriptase enzyme, the same enzymatic principle that has brought forth the domain Eukaryota. Free energy combined with empty space to create matter. More free energy combined again with matter to evolve it into more complex forms. The question then arises of the nature of the relationship between energy and matter. That is when thermodynamics comes in with one of its major teachings, that energy is convertible. Energy can be transacted into matter, as we have seen, in the presence of empty space. This principle is also expressed in the famous Einstein equation $E = mc^2$. The squared speed of EM radiation is a factor in the amount of empty space created with the disintegration of a certain mass. The convertibility of energy has been codified into the first law of thermodynamics and rather expressed by physicists as the principle of the conservation of energy. The first law of thermodynamics is considered a means to an end. In this context, it is ancillary to the second law. It is through the first law that the universe directly derived all its structures, but in the general framework of the quest for equilibrium as dictated by the second law.

The complexity of free energy is converted through chemical evolution into amino acids and also coded into bits of information represented by the bases A, G, C, and U to form ribonucleoprotein systems, the first configurations of life. Parts of the genomic language will, in time, be translated also into other complex structures, such as the cell membrane, the bacterial flagellum, and the eye, or into complex biochemical systems, such as the blood coagulation cascade or the Krebs cycle, etc. The genome is then a potential for complexity. In other words, complex living structures and systems of life represent phenotypic translations of the complexity and order of free energy as encoded in the genetic material. It is then a well-established principle that complexity can be built up according to the arrow of time, contingent upon the availability of free energy.

We contend that life evolved by the incorporation into its structure of free energy converted into new genetic information. Different mechanisms underlie this genomic size increase: rolling circle mechanisms of viroids, uptake of cellular genes by viruses, different methods of horizontal genetic transfer in bacteria, etc. In the case that concerns us directly here—that is, the evolution of eukaryotes—it is mediated by the copy and paste mechanism of retrotransposons. The size increase of the genome affords it an increase in its thermodynamic potential that is also reflected in the phenotype of the organism. Evolution operates through the direct creative power of free energy and has for impetus the imperative to increase the entropy of the universe. It resulted in the creation of energy-processing structures, which are themselves islands of increased order and complexity in the ocean of increasing entropy and disorder of the overall universe. Life feeds on free energy to forestall decay, multiply, and evolve. In the context of eukaryotic evolution, the configuration A of a lineage undergoes a modification of the mode of expression of its genes and accumulates information through retrotransposition or gene duplication, resulting in an increase of thermodynamic potential from configuration A to configuration B.

However, again, this is where the first and second laws of thermodynamics interact. The universe and its subsystems constitute dissipative structures, in the parlance of the Belgian physical chemist and Nobel laureate Ilya Prigogine.[2] The Big Bang is a manifestation of the law of universal radiation, according to which any structure with a temperature above absolute zero emits radiation in order to reach equilibrium. The entropy of an isolated system such as the universe continues on an ascending slope toward equilibrium. However, the continuous feeding on free energy of an open system such as life can temporarily dwarf the effect of the second law, allowing an entity to develop and evolve. It is in this context that the thermodynamic potential of a lineage increases as it evolves or changes configuration. We define the thermodynamic potential of a group of organisms as the amount of entropy they would produce for the environment during their natural* lifetime and life experiences. How then do we define biological evolution?

It is the neo-configuration of a biological system that accompanies an increase in its thermodynamic entropy output while its internal entropy decreases. It may be less complicated to estimate the entropy production of a biological system, or any open system for that matter, than its internal entropy. The evolution of a biological lineage from phenotype A to phenotype B is rendered by the entropy equation $E_B = xE_A$, where x, the entropy production ratio, is > 1. E represents the average entropy output and is inversely proportional to the system's internal entropy. This equation can also be formulated as follows: $\frac{E_B}{E_A} > 1$.

The differential entropy output between a certain configuration and any of its predecessors (ΔE) always increases:

$$\Delta E = E_B - E_A > 0$$

Those are the principles that have guided biological evolution. Viewed from the angle of the energetic requirements of a particular lineage, biological evolution is concerned with the crafting of more and more inefficient entities regarding life but more efficient in entropy production. The *Homo sapiens* individual who lives in the 21st century, for instance, produces on average more entropy for the universe than a *Homo erectus* counterpart who lived the same duration. It is also noticeable that within a same snapshot of time, a similar comparison can be made between people of different cultures and backgrounds, say between the Western man and the aborigine of the Amazonian Forest or the man of the OvaHimba

* That is, not tampered with by the disruptive actions of mankind.

tribe of Southwest Africa. The different “human races”, shaped by geography, do not necessarily occupy the same position or abscissa on the entropy production chain. Cultural differences amongst groups of people around the world constitute a major determinant of this disparity, which also explains the contrast between the civilizations they have built.

In summary, all structures in the universe answer to the requirement of the second law of thermodynamics, which is to increase the entropy of the global environment. Life is no exception. While the planets, stars, and galaxies hardly vary in this endeavor after a long period of construction, life, for its part, is in perpetual construction or evolution. Hence, its rate of entropy production increases with time.

Evolution by Thermodynamic Guidance in Action!

However, there may be no better illustration of this entropy equation than the concordance of the evolution of the eukaryotic lineages with their progressive capacity to increase the entropy of the universe. Biological species living across a snapshot of time do not result from the brutal selection among randomly evolved competing candidates according to a survival-of-the-fittest model, but rather constitute categorical imperatives of time-dependent thermodynamic constraints.

We call evolution by thermodynamic guidance the orthogenetic march of the lineages from one configuration to the next, the latter being more effective than the former at increasing the entropy of the universe. The theory of evolution by thermodynamic guidance is defined more by the consequence of the process of biological evolution for the universe in general than for the lineages themselves. Reverse thermodynamics would be a more appropriate concept for the lineages themselves, as they tend to increase in order or to accrue more intrinsic negative entropy as they evolve. We have elected to define evolution more in terms of the thermodynamic participation of the lineages in the cumulative entropy of the universe.* This theory of thermodynamic evolution would be falsified if it could be demonstrated that the configuration B of a lineage that has evolved from its configuration A has a slower metabolism and is less performant at increasing the overall entropy of the universe than any of its previous prototypes. Cases of prolonged sporulation, hibernation, or permafrost preservation, of course, are obviously not considered evolutionary stages.

The expectation would be that as the lineages evolve, they would become more efficient at converting highly ordered energy to lower-frequency radiation. Sponges constitute one of the earliest configurations of animals. The entropy they produced is almost equal to their basal metabolic rate. Later, the unambiguous expression of bilateral symmetry will allow some specific lineages to transition from a passive to an active mode of existence.[3] Bilateral symmetry allows movements such as swimming, creeping, or burrowing, thus recruiting the new configurations for a more robust level on the entropy creation scale. Even though these motor innovations afford the evolved organisms better aptitudes in getting food, escaping danger, choosing mates, or finding more suitable environments, they were essentially geared toward upscaling entropy rather than being just mere adaptations.

The linear progression of certain vertebrates from the phenotypes of fish to those of birds or mammals, passing through the successive configurations of amphibians, reptiles or synapsids, is consistent with an overall evolution of some ectothermic animals (known as "cold-blooded") to endothermic (so-called "warm-blooded") animals. Organisms said to be ectothermic have a basal metabolic heat

---

* At least for the reason stated above, that the entropy output of a system is easier to quantify than its internal entropy.

production too low to significantly affect their body temperature. Consequently, the heat that determines their body temperature comes almost entirely from the environment. Conversely, endothermic animals derive the heat of their body temperature largely from their own metabolism.[4, 5] It is, however, important to realize that the metabolism of all living organisms, including the so-called cold-blooded animals, produces a certain amount of endogenous heat. It is only that the latter group needs an additional or external source of heat to keep a functioning body temperature. In fact, it has been documented that some species of python can increase their body temperature by physiological means such as shivering (facultative endothermy) in order to keep their eggs warm.[6]

Besides, universal radiation is mandated by the second law of thermodynamics until equilibrium or the temperature of 0 K is reached for the whole universe. The evolution of vertebrates from ectothermy to endothermy would best be demonstrated, for instance, by evaluating the basal metabolism and activity patterns of an avian or mammal lineage over the course of many successive configurations. It is obviously not entirely possible due to the limited set of data one can glean in that regard from the fossil record. In general, fish, amphibians, and reptiles are ectothermic, while endothermy constitutes the bioenergetic system of most birds and mammals. Obviously, the metabolism of those different classes of animals is more complicated than this gross characterization, but the previous statement reflects the general pattern. We believe that the evolution of any lineage constitutes a gradual demarcation away from ectothermy. In that vein, the bioenergetic processes of the dinosaurs can be situated on a line between the metabolism of their amphibian ancestors and that of their neo-configurations, including their avian descendants, a statement that falls in line with the characterization by Grady and collaborators of the metabolic rates of the dinosaurs as intermediate between those of ectotherms and endotherms.[7]

According to our model, the thermoregulatory aspect represents more of a corollary than a purpose of evolution, particularly if one considers that endothermy evolved during the Mesozoic era when the Earth was overall warmer than it is today.[8] The evolution of the lineages is rather driven toward greater efficiency at increasing the entropy of the universe. In that regard, a certain number of biological modifications have accompanied the transition from ectothermy to endothermy, affording the endotherms with a greater metabolic rate. Ectotherms are more reliant upon anaerobic metabolism during activity. This form of respiration is not known as the most efficient modality of energy production. For instance, compared to aerobic glycolysis, the anaerobic metabolism of glucose produces sixteen times less ATP.[9] By contrast, endotherms have enhanced aerobic capacities, including a higher concentration of mitochondria in their tissues. The levels of oxygen consumption of endotherms both at rest and during activity exceed those of ectotherms by an average magnitude of five- to tenfold.[8] Furthermore, mammals process food ten times faster than reptiles on account of a greater intestinal surface area, a faster transit time through the gut, and a greater extraction efficiency.[10]

A greater fraction of energy consumed by endothermic animals, such as mammals, is devoted to metabolism and the production of heat as compared to ectothermic animals. Donald Prothero and Robert Dott Jr., in the eighth edition of their book *Evolution of the Earth*, express the fact that 88% of the food consumed by a lion, a mammal, goes directly to producing atmospheric entropy and is lost as nonusable heat, compared to 70% in a crocodile, a reptile.[11] As a result, a mammal or bird requires from 5 up to 20 times more food for its maintenance than a similar-sized reptile or other vertebrate ectotherm.[8, 10, 12]

Moreover, the principle of thermodynamic evolution is consistent with the second law of thermodynamics, whereby the entropy of the universe increases according to a time arrow. Working in consonance with this principle, some female birds act as gatekeepers of the evolution of their lineage. The mate choice of some female birds is instructed by the degree of fitness of the male as revealed in his courtship display. For the female, "beauty" or intrinsic order is defined as the capacity of the male to increase the entropy of the universe compared to its rivals and is adjudicated on the intrinsic or local order he can display. The beauty of the peacock's tail, for example, is a surrogate for this very feature. Such is the nature of the mate assessment carried out, for instance, by the peahen, the female bowerbird, or the female bird of paradise. It is the same selection method used by females of numerous species that enjoy the prerogative to choose their mate among competing males. In human societies, the display of wealth by men or of beauty by women pertains to the same logic. The quest for an increasing entropy of the universe constitutes the pressure under which phenotypes evolve.

The Special Case of Mankind

The human lineage has evolved along the same bioenergetic line as all other mammals, with our own basal metabolic rate and activity patterns, of course. But the story of our civilizations is one of the things that sets us apart from other animals. The history of humanity can be traced along the lines of a steady increase in energy utilization and technological innovations on par with a similar pattern in our rate of entropy production for the universe. The crafting by our *Homo erectus* ancestors of the first hand axe approximately 2 million years ago in the surroundings of modern Kenya to facilitate among other uses the extraction of game meat, as well as their unprecedented control of fire, again in Africa over a million years ago,[13-15] proceed from the same drive as any major technological innovations of modern times. Darwin himself opined that, besides language, the mastery of fire probably ranked as the greatest achievement of humanity.[16] A modus vivendi of hunting and gathering yielded in time to a more sedentary lifestyle through the Neolithic agricultural revolution. About 12,500 years ago, humans took it upon themselves to produce and multiply their own food. However, through the use of force, some will gather the fruits of the labor of others. Slavery, one of the earliest modes of energy harnessing, will not be outlawed in the United States, for instance, until the second half of the 19th century, from its apogee a century earlier.[17]

Progressively, however, new energy sources from animal traction, wind, and water would be domesticated to supplement manual labor. Later, the Industrial Revolution, which started in the late 18th century, would gradually replace a system of production based primarily on manpower with one where mechanized factories played a critical role. This new era owed a major debt to the invention and development of the steam engine, which would eventually be followed by more advanced technologies, until the information revolution of our time. Along the way, we have also developed the know-how to exploit other major energetic modalities, from fossil fuels to alternative energy sources.

This small historical snapshot clearly displays our tendency for the use of a growing amount of energy, all the while increasing the entropy of our environment. Indeed, we were all built for this purpose. The human brain, which epitomizes one of our major differences with the other lineages, and which is the major factor of our dominance in nature, uses one-fifth of the body's energy regardless of its level of activity.[18] The erection of ordered structures such as the Stonehenge monument in England, the north African pyramids, the Incan Machu Picchu, or our more modern engineering and technological accomplishments, arguably partake of the same vocation of processing free energy to increase the overall disorder of the universe by creating order locally. The differences in energy utilization and entropy production across cultures conceivably result from an unequal disposition toward the creation of a certain level of entropy for the universe. Indeed, although we are all built to process energy, we are determined to do so at different levels, as an examination of specific snapshots of human history reveals. The differences amongst cultures are rather ascertained on an entropy production scale. They boil down to the question of which cultures create more or less entropy for the universe.

Where Do We Go from Here?

It is expected that our lineage will evolve the skills to tap deeper into the diverse modalities of energy reserves of the universe and make use of them. This path, however, will result in the creation of more and more entropy for the universe. In fact, the sum of human interactions with the physical environment does not form a negative or Lovelockian feedback loop in the sense of a continuous, self-regulating ecosystem. Consequently, we need a modified Gaia concept; one that considers our planet as a place, though predestined for the emergence of life, we will need to abandon in order for us to survive, because of the negative impact of our actions. In reality, however, we are not to blame; it was bound to happen.

The global human population has been increasing steadily since time immemorial.[19] Likewise, carbon dioxide concentrations in the Earth's atmosphere have followed an upward trend at least since measurements made in the 1960s.[20] This latter circumstance, certainly linked to the former, will undoubtedly lead to an increasing Earth temperature and other climate changes negatively affecting life. The varied effects of the increased heat resulting from our activities constitute what is known as anthropogenic global climate change. We can try to delay this process, but it is inexorable. Those are some of the considerations that have led us to think that our cradle, Mother Earth, for beautiful as it is, is not our destination.

Our future is out of the closed planetary constraints of the Earth, in the open universe where free energy is plentiful and where there are no practical limits to the amount of entropy we can produce. It is not far-fetched to contend that we will press on our path of activities against different levels of gravity,* whether they involve our planet, an asteroid, a star, or a black hole. With the help of our space programs, we have made a few cracks through the shell of our egg, Mother Earth. We have taken a few human steps outside but cannot yet venture too far.** It will take us time to perfect our skills, but space is the place of our future. Our destiny is in the skies.

---

* Prime opportunities to produce entropy.

** We have been farther with our machines.

Epilogue

In this project, we have identified the second law of thermodynamics as the law par excellence of the universe. A great many scientists—the likes of Nicolas Carnot, Rudolph Clausius, Lord Kelvin (William Thomson), James Maxwell, Ludwig Boltzmann, Josiah Gibbs, and Claude Shannon—are associated with its discovery and characterization. Directly or indirectly, the second law is involved in all the phenomena that have derived from the Big Bang singularity, including the appearance and evolution of matter, inert and living alike. Our proposition that a physical law governs the phenomenon of life and its evolution is obviously not new. Richard Owen, for instance, a naturalist of Victorian England, suggested that life was submitted to some form of natural law.[1] But how are we to make sense of the emergence of life and particularly of humans in the history of the cosmos? We can start by emphasizing our belief that energy, space, and eternity (the fullness of time)—three inseparable entities—constitute the fundamental features of the universe. We then submit that the initial conditions of the universe are the values of the Planck units associated with these fundamentals. We suggest that it is the relationships between those values that determine the universe's physical constants such as the speed of light, the gravitational constant, or the Planck's constant, etc.*

---

* One can always wonder about the origin of the respective values of the Planck units, but the speed of light is obtained by dividing the Planck length by the Planck time. Likewise, the gravitational constant G is given by the ratio of the cube of the Planck length to the Planck mass divided by the square of the Planck time. In the same spirit, the reduced Planck's constant in joule-second is roughly the number obtained by dividing the Planck energy by the value of one second

These relationships also determine the future of the universe from Big Bang to thermodynamic equilibrium. The future determined by these initial conditions and relationships is the universe as we know it, including us humans, whose skills in creating entropy are quite evident and expand with time. The universe is not, strictly speaking, fine-tuned for life;* rather, it is geared toward equilibrium, and the phenomenon of life finds itself right in the middle of that process playing its part. The universe was inexorably bound to create the conditions for the emergence of both inert matter and living beings as mediators and agents for the ultimate goal of equilibrium. On an anthropic scale, this is also, in our opinion, the answer to the Gouldian "question of the ages- Why do humans exist?",[3] rather than the supposed fortuitous sparing of a remote chordate ancestor named *Pikaia gracilens* from the so-called Cambrian- Ordovician extinction. Rewind the universe back to the Big Bang singularity and play it again, 13.8 billion years later, we would be exactly where we are today.

Obviously to study life in depth, one has to walk upstream and start at the Big Bang, the event that unleashed the energy of the universe and the second law of thermodynamics. The importance of this law in the phenomena studied by physicists is quite well known, but its interaction with life systems, despite the brilliant *What is Life* essay of Erwin Schrödinger, has not been thus far clarified in a systematic and meaningful way. Still, under the impulse of the second law, life appeared and evolved through the ancillary principle of energy convertibility of the first law.

---

in Planck time. In other words, the Planck dimensions are fundamental, and it is from their relationships that the aforementioned constants arose, even though those constants were known first and that it was from them that Max Planck derived some of the units that bear his name.[2]

* The view that the universe is fine-tuned for life can be considered as the biocentric principle, or some form of biocentrism, an extension of the anthropic principle.

One special step in the evolution of life was the fusion of certain bacteriophages with their bacterial hosts to form the eukaryotes. This form of genetic interaction, heralded in this context and in all likelihood by viroid infection of RNA cells, constitutes the origin of all modalities of fertilization, including sex. Animal cells are generally diploid,* while prokaryotes are haploid. Retrospectively, one can understand how the eukaryotic diploidy has resulted from the merging of two haploid entities, a virus and a bacterium, in the same manner that generally a eukaryotic organism originates from the merging of a sperm and an ovum.

We have also shown that the eukaryotic lineages are fixed. They do not split. Our common ancestry with the bonobo is not among the eukaryotes. It is, on one side of the bilateral descent system, a retrovirus that subsequently diverged by the agency of an RNA polymerase enzyme. On the other side, it is the *Thiomargarita namibiensis* bacterium. The diagram of eukaryotic life is the transposition to a new domain of a tree of retroviral quasispecies, thus fundamentally of RNA genome at its roots. (See chapter 5, figure 8).

And, if we take appropriate stock of our viral origin, we will understand that biodiversity, which initially stems from RNA polymerase enzymes, rather than a liability, is a strength built into the fabric of life. The Vignuzzi study has shown that a viral clone is more likely to succeed at reaching its goal and multiply when backed up by a diverse rather than by a genetically constrained population.[4,5] The central concept of the quasispecies theory and consequently of the diversity of life is that the success of a population occurs through the complementing functions of individual variants and different subpopulations, rather than by the exclusive selection of highly performant mutants. A certain

* Notable exceptions are gamete cells and cells of male insects of the order Hymenoptera which are haploid. Additionally, polyploidy, in which the homologous chromosomes of the cells occur in sets of more than two units, is common in plants and fungi.

level of diversity is a strength rather than a weakness. The lesson for us humans is that for as different our civilizations and levels of technological achievements may seem across the planet, our success as a lineage is through cooperative interactions rather than through zero-sum game competition or a survival-of-the-fittest framework. This perspective can be analogized to the success of the Apollo 11 mission, which led to the first landing of our species on the moon in July 1969.

Neil Armstrong, Edwin Aldrin, and Michael Collins were exceptionally talented astronauts who deserved all the praise and credit for their role in that proverbial "giant leap for mankind." However, by themselves, alone or cloned to the thousands, they would not have been able to successfully complete such a mission. This project needed people with different sets of skills, from planning to execution, including scientists, engineers, flight controllers, different types of technicians, and last but not least, President Kennedy with his bold objective to "land . . . a man on the moon and return . . . him safely to the earth."[6] It was a mission of cooperation between complementing skills. Further in his speech, the president also acknowledged this much when he cautioned: "But in a very real sense, it will not be one man going to the moon—if we make this judgment affirmatively, it will be an entire nation. For all of us must work to put him there."[6] In the Vignuzzi study,[4] a clone of neurovirulent poliovirus was able to cross the blood-brain barrier of the mice and replicate themselves only when they were supported by a diverse viral population. This framework runs counter to the ferocious Spencer-Darwinian model summarized in the concept "survival of the fittest," where one needs to keep a constant watch to avoid exploitation or predation. A framework of cooperation and interdependence is more aligned with our quasispecies origin than the brutal, tooth-and-claw struggle of social Darwinism, which divides members of society between victors and vanquished. Indeed, even at the height of the fervor of the rugged individualist philosophy of the late 19th and early 20th century, dissenting voices had started to emerge. Among them was the Russian naturalist and philosopher Peter Kropotkin who, in his book *Mutual Aid: A Factor of Evolution*, published in 1902, brought up evidence of widespread cooperation in nature, especially among conspecific animals.[7] The late biologist Lynn Margulis was one of the most recent vocal opponents of the view of biological evolution as the result of the triumph of the fittest in some kind of universal warfare for existence. And obviously, even though we are not a proponent of her version of the endosymbiotic theory of eukaryogenesis, we support her contention that it was due to cooperation instead of competition that life was able to spread to the far reaches of the world.[8] Undoubtedly, warfare within and without the species is part of the reality of nature. It can lead to the replacement of one group or lineage in an ecosystem by another but does little to advance life itself into hitherto uninhabited niches.

But to return to the general law of the universe, we present thermodynamic evolution as a new paradigm for the natural sciences. The purpose of all matter is to process into disordered heat the energy of the universe. Electromagnetic radiation seems to have emerged from nothingness only to harken back to it at a time determined by its speed. This is how the notion of free will goes up in smoke. In truth, what does this feature of electromagnetic radiation imply for this gift that is dear to us, our supposed free will, if we do not have a choice in the matter? Our evolution from *Australopithecus* to fire-lighting *Homo erectus* was built in and timed. So were the Industrial Revolution and our developing conquest of space. The story of tomorrow may be printed somewhere on some hidden stone.

However, as for me, illusory or not, at the moment of my choosing between A and B, I am conscious of making a choice, and that is good enough. And, if I may be allowed an opinion, I think that should be good enough for us humans. It might also be that while the collective choice is predetermined, my free choosing of B might compel somebody else to choose A if it were also predetermined that both items were to be chosen. As if we were all involved in a scheme akin to quantum entanglement where, in the case of an entangled pair of particles, the measuring on a certain axis of the direction of the spin of one of them determines the other particle to spin in the reverse direction of that same axis. Nonetheless, the theories exposed in this work should not be used as an alibi to absolve anybody of their legal and moral obligations.

Entangled particles behave as if they are still part of a singularity whatever the distance that seems to separate them, and our comments in the following lines regarding the behavior of flatline radiation apply to them as well. In a singularity, the frequencies of radiation are superposed, as in one indivisible unit of time. Daughter particles resulting from the decay of an original particle with zero spin, even when separated, still coexist within the same unit of time. The measure of time given by the superposed radiation frequencies of the singularity does not change, even when the latter is stretched by the physical separation of the daughter particles. Indeed, the length of a spring, if determined by its number of coils, does not change whether the coils superpose (singularity) or are stretched. In other words, the real distance that separates the two daughter particles is the one given by the time that separates them, which is none, but not the distance measured by an external, isolated ruler.

The reader must have realized that this book, rather than a monograph, is a compendium of several theses dealing with the origin of living systems from viroids, to viruses, prokaryotes, and eukaryotes. Those propositions, however, can all be organized under the seeds-first theory of eukaryotic biodiversity. Indeed, viroids, viruses along with bacteria are all positioned on the assembly line leading to the original eukaryotic zygotes.

The book also presents the universe as organized energy in search of equilibrium, creating and evolving matter in the process. Life then, like all matter, is an epiphenomenon, an intermediary state between the organized energy of the Big Bang and the flat radiation of thermodynamic equilibrium.

Furthermore, we threw a new light on our interaction with our common home, Planet Earth. Just as eggs or bacteria are fertilized by sperm or bacteriophages, respectively, we explain our presence on Earth through its fertilization by late heavy bombardment (LHB) meteorites about 4 billion years ago. We contend that those meteorites impregnated our planet with life through both the direct and indirect supply of prebiotic compounds,[9-12] such as nucleobases, amino acids, and their precursors, etc. The earliest life forms spearheaded the emergence of photosynthetic organisms that will in turn generate the conditions of the appearance of the earth fauna, including our ancestors, through the Great Oxidation Event. But now that our impact on the planet is rendering it inhospitable, and our technological achievements are taking on the next challenge,* we are getting ready to escape this once comfortable home like a brood of bacteriophages deserting a lysed bacterium.

And to the question of whether we humans are special, whether we stand at "the very summit of the organic scale,"[13] it may seem rather difficult to answer in the affirmative when we take stock of our vulnerability, when we consider, for instance, afflictions such as the 14th-century bubonic plague** or the 1918 flu pandemic that humanity experienced at the hands of minuscule life forms invisible to the naked eye.

As of this writing, severe acute respiratory syndrome coronavirus 2 (SARS-CoV-2), the strain of coronavirus that causes the coronavirus disease 2019 (COVID-19) pandemic, is driving us off

---

* Leave Earth.

** Bubonic plague was caused by Yersinia pestis, a flea-borne bacterial disease of rats that spread to humans.

our streets, restaurants, and workplaces. Under these conditions, it may seem unreasonable, once again, to portray humankind as the ne plus ultra of creation. But on the other hand, it can be argued that, on balance, there is something unique about the human condition, that our special status comes from the great responsibility that falls to us given our ability to influence the fate of all life on Earth, whether through genetic engineering technologies or the changes we are creating in the climate.

The Universe is the Illusory Unfolding of Eternity

By following the predicted circular path of the universe to thermodynamic equilibrium we were able to appreciate some of the attributes of Nothingness. We have demonstrated that Nothingness has structure and is describable. It is a physical necessity found at the same time at the beginning and at the end of the universe. It is both the genitor of the universe and its final destiny. Nothingness is completely uniform; it has superlative entropy. Swapping the position of different parts of the system will not change its configuration. Nothingness is flatline electromagnetic radiation* filling the entirety of the universe in equilibrium. It is the total possible space and the total possible time in one single instant: the Now. Expressed differently, it is the manifestation all at once (*totum simul*) of the whole of spacetime. There is no beyond; there is no past and no future. It does not age, as it knows no flow of time. In short, Nothingness is eternal.** Nothingness is God.

---

* That is radiation of zero frequency and infinite wavelength. Flatline radiation is instantaneously here and everywhere.

** Eternity in the sense of Boethius and Thomas Aquinas according to which the fullness of time is experienced all at once.[14] A view that echoes Plato's famous description of time as "a moving image of eternity."[15]

Indeed, those attributes of Nothingness worked out by reason and science align, we believe, quite well with the biblical God and in some ways with the God alluded or referred to by Parmenides and Plato. The following quote from the poem of Parmenides seems to encompass the characteristics of Nothingness, and of course of God: "We can speak and think only of what exists. And what exists is uncreated and imperishable for it is whole and unchanging and complete. It was not or nor shall be different since it is now, all at once, one and continuous." The same can be said of a description of *the father and creator*'s relationship with time in the *Timaeus* of Plato:[15] ". . . for we say that he 'was,' he 'is,' he 'will be,' but the truth is that 'is' alone is properly attributed to him." In the *Confessions* of St. Augustine,[16] we read the same appraisal of God's eternity: "Your years neither go nor come, but our years pass and others come after them, so that all may come in their turn. Your years are completely present to you all at once, because they are at a permanent standstill." Those passages meet the self-description that God gave to Moses in Exodus 3:14: "God said to Moses, 'I AM WHO AM.' He said: 'Thus shall you say to the sons of Israel: HE WHO IS has sent me to you.'"

On this issue, since the prophet Isaiah was among those who spoke first, we deem it proper to refer to his writings for the last word. And we have found it in the 6th verse of the 44th chapter: "Thus saith the LORD the King of Israel, and his redeemer the LORD of hosts; I am the first, and I am the last; and beside me there is no God."

Acknowledgments

I am indebted to my wife, Mich, for her support, her patience, and for being my subtle cheerleader while I was working on the book, my daughters, Reggie and Jassie, for having provided my initial inspiration for it, and for never failing to remind me when I have violated my own deadlines, which was often the case. In more ways than one, you have all been a part of this journey. Additionally, I express my sincere gratitude to the staff at Gatekeeper Press for their assistance in transforming the original manuscript into this final product. I particularly appreciate the expertise of editors Jason Pettus and Taleiah Todd-Hill, cover artist Haroula Kontorousi, graphic artist Amna Ahsan, and author manager Kelly Santaguida. The first editing work was done by Jason Pettus, and I was amazed at all the inputs he provided, including helping me translate my overlooked Gallicisms into proper English, and other valuable suggestions. The mistakes that remain are my own only. I have no ground whatsoever to be stubborn, especially when it comes to a language in which I am only half-fluent, if that. Let's only hope that time will bring wisdom.

List of Figures and Tables

Chapter 1

Chapter 2

Chapter 3

Chapter 4

Chapter 5

Chapter 6

Chapter 9

Chapter 10

Chapter 13

Chapter 14

Chapter 15

Bibliography

General References

1. Acheson, Nicholas H. *Fundamentals of Molecular Virology*. Wiley, 2011.
2. Alberts, Bruce. "Molecular Biology of the Cell." *National Center for Biotechnology Information*, U.S. National Library of Medicine, 1 Jan. 1970. https://ncbi.nlm.nih.gov/books/NBK21054/
3. Calendar, Richard. *The Bacteriophages*. Oxford University Press, 2006.
4. Cordingley, Michael G. *Viruses: Agents of Evolutionary Invention*. Harvard University Press, 2017.
5. Engleberg, N. Cary, et al. *Schaechter's Mechanisms of Microbial Disease*. Wolters Kluwer Health/Lippincott Williams & Wilkins, 2013.
6. Gilbert, Scott F. "Developmental Biology." *National Center for Biotechnology Information*, U.S. National Library of Medicine, 1 Jan. 1970. https://ncbi.nlm.nih.gov/books/NBK9983/
7. Lodish, Harvey. *Molecular Cell Biology*. W. H. Freeman and Co., 2016.
8. Murray, Patrick R., et al. *Medical Microbiology*. Elsevier, 2016.
9. Tennant, Paula, et al. *Viruses: Molecular Biology, Host Interactions and Applications to Biotechnology*. Academic Press, 2018.
10. Voet, Donald, et al. *Fundamentals of Biochemistry: Life at the Molecular Level*. J. Wiley & Sons, 2016.

Preface

1. Einstein, Albert. La Relativité: Théorie De La Relativité Restreinte Et Générale: La Relativité Et Le Problème De L'espace. Éditions Payot & Rivages, 2001.
2. Darwin, Francis & Seward, A. C. eds. 1903. *More letters of Charles Darwin. A record of his work in a series of hitherto unpublished letters.*

London: John Murray. Volume 1, p. 209. (http://darwin-online.org.uk/content/frameset?itemID=F1548.1&viewtype=text&pageseq=1)

Chapter 1

1. NASA. "About Life Detection." *NASA*, NASA. https://astrobiology.nasa.gov/research/life-detection/about/. Accessed online on 4/17/18
2. Bhattacharyya, T., et al. "Mechanistic Basis of Infertility of Mouse Intersubspecific Hybrids." *Proceedings of the National Academy of Sciences*, vol. 110, no. 6, 2013, doi:10.1073/pnas.1219126110.
3. Maheshwari, Shamoni, and Daniel A. Barbash. "The Genetics of Hybrid Incompatibilities." *Annual Review of Genetics*, vol. 45, no. 1, 2011, pp. 331–355., doi:10.1146/annurev-genet-110410-132514.
4. Johnson, Norman A. "Hybrid Incompatibility and Speciation." *Nature News*, Nature Publishing Group, 2008. https://nature.com/scitable/topicpage/hybrid-incompatibility-and-speciation-820/. Accessed online on 4/17/18.
5. Lincoln, T. A., and G. F. Joyce. "Self-Sustained Replication of an RNA Enzyme." *Science*, vol. 323, no. 5918, 2009, pp. 1229–1232., doi:10.1126/science.1167856.
6. International Human Genome Sequencing Consortium. Initial se8quencing and analysis of the human genome. *Nature* 409, 860–921 (2001). https://doi.org/10.1038/35057062
7. Moelling, Karin. *Viruses: More Friends than Foes*. World Scientific, 2017, p. 62.
8. Moelling, Karin, et al. "RNase H as Gene Modifier, Driver of Evolution and Antiviral Defense." *Frontiers in Microbiology*, vol. 8, 2017, doi:10.3389/fmicb.2017.01745.
9. Cochrane, Alan. "Retroviruses." *Fundamentals of Molecular Virology*, by Nicholas H. Acheson, Wiley, 2011, p. 344.
10. Folimonova, S. Y. "Superinfection Exclusion Is an Active Virus-Controlled Function That Requires a Specific Viral Protein." *Journal of Virology*, vol. 86, no. 10, 2012, pp. 5554–5561., doi:10.1128/jvi.00310-12.
11. Michel, Nico, et al. "The NEF Protein of Human Immunodeficiency Virus Establishes Superinfection Immunity by a Dual Strategy to

Downregulate Cell-Surface CCR5 and CD4." *Current Biology*, vol. 15, no. 8, 2005, pp. 714–723., doi:10.1016/j.cub.2005.02.058.

12. Leiman, P. G., et al. "Structure and Morphogenesis of Bacteriophage T4." *Cellular and Molecular Life Sciences (CMLS)*, vol. 60, no. 11, 2003, pp. 2356–2370., doi:10.1007/s00018-003-3072-1.
13. Ali, Yahya, et al. "Temperate Streptococcus Thermophilus Phages Expressing Superinfection Exclusion Proteins of the LTP Type." *Frontiers in Microbiology*, vol. 5, 2014, doi:10.3389/fmicb.2014.00098.
14. Clift, Dean, and Melina Schuh. "Restarting Life: Fertilization and the Transition from Meiosis to Mitosis." *Nature Reviews Molecular Cell Biology*, vol. 14, no. 9, 2013, pp. 549–562., doi:10.1038/nrm3643.
15. Burkart, Anna D., et al. "Ovastacin, a Cortical Granule Protease, Cleaves ZP2 in the Zona Pellucida to Prevent Polyspermy." *Journal of Cell Biology*, vol. 197, no. 1, 2012, pp. 37–44., doi:10.1083/jcb.201112094.
16. Gilbert, Scott F. "Oogenesis." *Developmental Biology. 6th Edition.*, U.S. National Library of Medicine, 1 Jan. 1970. https://ncbi.nlm.nih.gov/books/NBK10008/
17. Schulz, H. N., et al. "Dense Populations of a Giant Sulfur Bacterium in Namibian Shelf Sediments." *Science*, American Association for the Advancement of Science, 16 Apr. 1999. https://science.sciencemag.org/content/284/5413/493.full
18. Sayama, Mikio. "Presence of Nitrate-Accumulating Sulfur Bacteria and Their Influence on Nitrogen Cycling in a Shallow Coastal Marine Sediment." *Applied and Environmental Microbiology*, vol. 67, no. 8, 2001, pp. 3481–3487., doi:10.1128/aem.67.8.3481-3487.2001.
19. Kojima, Hisaya, et al. "Ecophysiology of Thioploca Ingrica as Revealed by the Complete Genome Sequence Supplemented with Proteomic Evidence." *The ISME Journal*, vol. 9, no. 5, 2014, pp. 1166–1176., doi:10.1038/ismej.2014.209.
20. Nelson, D. C., et al. "Occurence and Regulation of Calvin Cycle Enzymes in Non-Autotrophic Beggiatoa Strains." *Archives of Microbiology*, vol. 151, no. 1, 1988, pp. 15–19., doi:10.1007/bf00444662.

21. Jannasch, Holger W., et al. "Massive Natural Occurrence of Unusually Large Bacteria (Beggiatoa Sp.) at a Hydrothermal Deep-Sea Vent Site." *Nature*, vol. 342, no. 6251, 1989, pp. 834–836., doi:10.1038/342834a0.
22. Phillips, David M, and Gil L Dryden. "Comparative Morphology of Mammalian Gametes." *A Comparative Overview of Mammalian Fertilization*, by Bonnie Dunbar and Michael O'Rand, Springer Science+Business Media, 1991, p. 44.
23. Cooper, Geoffrey M. "Meiosis and Fertilization." *The Cell: A Molecular Approach. 2nd Edition.*, U.S. National Library of Medicine, 1 Jan. 1970. https://ncbi.nlm.nih.gov/books/NBK9901/
24. Alberts, Bruce. "Eggs." *Molecular Biology of the Cell. 4th Edition.*, U.S. National Library of Medicine, 1 Jan. 1970, www.ncbi.nlm.nih.gov/books/NBK26842.
25. Austin, C R. "Evolution of Human Gametes-Oocytes." *Gametes*, by John L. Yovich and Jurgis Gediminas Grudzinskas, Cambridge University Press, 1995, p. 13.
26. *Nature News*, Nature Publishing Group. https://nature.com/scitable/definition/gamete-gametes-311/. Accessed online on 9/1/2021.
27. Pansky, Ben. "Gametogenesis: Oogenesis." *LifeMap Discovery®*. https://discovery.lifemapsc.com/library/review-of-medical-embryology/chapter-4-gametogenesis-oogenesis. Accessed online 8/29/21.
28. Schulz, Heide N., and Dirk De Beer. "Uptake Rates of Oxygen and Sulfide Measured with Individual Thiomargarita Namibiensis Cells by Using Microelectrodes." *Applied and Environmental Microbiology*, vol. 68, no. 11, 2002, pp. 5746–5749., doi:10.1128/aem.68.11.5746-5749.2002.
29. Graco, Michelle, et al. "Massive Developments of Microbial Mats Following Phytoplankton Blooms in a Naturally Eutrophic Bay: Implications for Nitrogen Cycling." *Limnology and Oceanography*, vol. 46, no. 4, 2001, pp. 821–832., doi:10.4319/lo.2001.46.4.0821.
30. Zopfi, Jakob, et al. "Ecology of Thioploca Spp.: Nitrate and Sulfur Storage in Relation to Chemical Microgradients and Influence of Thioploca Spp. on the Sedimentary Nitrogen Cycle." *Applied and*

Environmental Microbiology, vol. 67, no. 12, 2001, pp. 5530–5537., doi:10.1128/aem.67.12.5530-5537.2001.

31. Fischer, B., and B. D. Bavister. "Oxygen Tension in the Oviduct and Uterus of Rhesus Monkeys, Hamsters and Rabbits." *Reproduction*, vol. 99, no. 2, 1993, pp. 673–679., doi:10.1530/jrf.0.0990673.
32. Yang, Yu et al. "Comparison of 2, 5, and 20% O2 on the development of post-thaw human embryos." *Journal of assisted reproduction and genetics* vol. 33,7 (2016): 919-27. doi:10.1007/s10815-016-0693-5
33. Bavister, Barry. "Oxygen Concentration and Preimplantation Development." *Reproductive BioMedicine Online*, vol. 9, no. 5, 2004, pp. 484–486., doi:10.1016/s1472-6483(10)61630-6.
34. Collado-Fernandez, Esther, et al. "Metabolism throughout Follicle and Oocyte Development in Mammals." *The International Journal of Developmental Biology*, vol. 56, no. 10-11-12, 2012, pp. 799–808., doi:10.1387/ijdb.120140ec.
35. Devreker, Fabienne, and Yvon Englert. "In Vitro Development and Metabolism of the Human Embryo up to the Blastocyst Stage." *European Journal of Obstetrics & Gynecology and Reproductive Biology*, vol. 92, no. 1, 2000, pp. 51–56., doi:10.1016/s0301-2115(00)00425-5.36.
36. Seli, Emre, et al. "Minireview: Metabolism of Female Reproduction: Regulatory Mechanisms and Clinical Implications." *Molecular Endocrinology*, vol. 28, no. 6, 2014, pp. 790–804., doi:10.1210/me.2013-1413.
37. Houghton, Franchesca D., et al. "Oxygen Consumption and Energy Metabolism of the Early Mouse Embryo." *Molecular Reproduction and Development*, vol. 44, no. 4, 1996, pp. 476–485., doi:10.1002/(sici)1098-2795(199608)44:4<476::aid-mrd7>3.0.co;2-i.
38. Dumoulin, J. C., et al. "Positive Effect of Taurine on Preimplantation Development of Mouse Embryos in Vitro." *Reproduction*, vol. 94, no. 2, 1992, pp. 373–380., doi:10.1530/jrf.0.0940373.
39. Devreker, F., et al. "Effects of Taurine on Human Embryo Development in Vitro." *Human Reproduction*, vol. 14, no. 9, 1999, pp. 2350–2356., doi:10.1093/humrep/14.9.2350.

40. Hignite Barr, Susan, and Gene Oliphant. "Sulfate Incorporation into Macromolecules Produced by Cultured Oviductal Epithelium." *Biology of Reproduction*, vol. 24, no. 4, 1981, pp. 852–858., doi:10.1095/biolreprod24.4.852.
41. Hyde, B. A., and D. L. Black. "Synthesis and Secretion of Sulphated Glycoproteins by Rabbit Oviduct Explants in Vitro." *Reproduction*, vol. 78, no. 1, 1986, pp. 83–91., doi:10.1530/jrf.0.0780083.
42. Lobo, Maria V.T., et al. "Immunohistochemical Localization of Taurine in the Rat Ovary, Oviduct, and Uterus." *Journal of Histochemistry & Cytochemistry*, vol. 49, no. 9, 2001, pp. 1133–1142., doi:10.1177/002215540104900907.
43. Battaglia, C., et al. "Embryonic Production of Nitric Oxide and Its Role in Implantation: A Pilot Study." *Journal of Assisted Reproduction and Genetics*, vol. 20, no. 11, 2003, pp. 449–454., doi:10.1023/b:jarg.0000006706.21588.0d.
44. Lipari, Christopher W., et al. "Nitric Oxide Metabolite Production in the Human Preimplantation Embryo and Successful Blastocyst Formation." *Fertility and Sterility*, vol. 91, no. 4, 2009, pp. 1316–1318., doi:10.1016/j.fertnstert.2008.01.108.
45. Nevoral, Jan, et al. "Endogenously Produced Hydrogen Sulfide Is Involved in Porcine Oocyte Maturation in Vitro." *Nitric Oxide*, vol. 51, 2015, pp. 24–35., doi:10.1016/j.niox.2015.09.007.
46. Li, Shuai, and Wipawee Winuthayanon. "Oviduct: Roles in Fertilization and Early Embryo Development." *Journal of Endocrinology*, vol. 232, no. 1, 2017, doi:10.1530/joe-16-0302.
47. Rosselli, M., et al. "Identification of Nitric Oxide Synthase in Human and Bovine Oviduct." *Molecular Human Reproduction*, vol. 2, no. 8, 1996, pp. 607–612., doi:10.1093/molehr/2.8.607.
48. Patel, Pushpa, et al. "The Endogenous Production of Hydrogen Sulphide in Intrauterine Tissues." *Reproductive Biology and Endocrinology*, vol. 7, no. 1, 2009, p. 10., doi:10.1186/1477-7827-7-10.
49. Ning, Nannan, et al. "Dysregulation of Hydrogen Sulphide Metabolism Impairs Oviductal Transport of Embryos." *Nature Communications*, vol. 5, no. 1, 2014, doi:10.1038/ncomms5107.

50. Kim, Bo Hyun, et al. "Involvement of Nitric Oxide During in Vitro Fertilization and Early Embryonic Development in Mice." *Archives of Pharmacal Research*, vol. 27, no. 1, 2004, pp. 86–93., doi:10.1007/bf02980052.
51. Sengupta, Jayasree, et al. "Nitric Oxide in Blastocyst Implantation in the Rhesus Monkey." *Reproduction*, vol. 130, no. 3, 2005, pp. 321–332., doi:10.1530/rep.1.00535.
52. Gouge, R. C., et al. "Nitric Oxide as a Regulator of Embryonic Development." *Biology of Reproduction*, vol. 58, no. 4, 1998, pp. 875–879., doi:10.1095/biolreprod58.4.875
53. Kolluru, Gopi K., et al. "A Tale of Two Gases: No and H2S, Foes or Friends for Life?" *Redox Biology*, vol. 1, no. 1, 2013, pp. 313–318., doi:10.1016/j.redox.2013.05.001.
54. Gardner, David K, et al. "Noninvasive Assessment of Human Embryo Nutrient Consumption as a Measure of Developmental Potential." *Fertility and Sterility*, vol. 76, no. 6, 2001, pp. 1175–1180., doi:10.1016/s0015-0282(01)02888-6.
55. Gardner, David K., and Michelle Lane. "Amino Acids and Ammonium Regulate Mouse Embryo Development in culture1." *Biology of Reproduction*, vol. 48, no. 2, 1993, pp. 377–385., doi:10.1095/biolreprod48.2.377.
56. Gardner, D K, et al. "Human and Mouse Embryonic Development, Metabolism and Gene Expression Are Altered by an Ammonium Gradient in Vitro." *REPRODUCTION*, vol. 146, no. 1, 2013, pp. 49–61., doi:10.1530/rep-12-0348.
57. Csonka, Frank A. "Nitrogen, Methionine and Cystine Content of Hen's Eggs. Their Distribution in the Egg White and Yolk." *The Journal of Nutrition*, vol. 42, no. 3, 1950, pp. 443–451., doi:10.1093/jn/42.3.443. 58.
58. Csonka, Frank A., et al. "The Methionine and Cystine Content of Hen's Eggs." *Journal of Biological Chemistry*, vol. 169, no. 2, 1947, pp. 259–265., doi:10.1016/s0021-9258(17)35023-8.
59. Jönsson, L., et al. "Production and Egg Quality in Layers Fed Organic Diets with Mussel Meal." *Animal*, vol. 5, no. 3, 2011, pp. 387–393., doi:10.1017/s1751731110001977.

60. Wang, Fu-Rong, et al. "Effects of Dietary Taurine on Egg Production, Egg Quality and Cholesterol Levels in Japanese Quail." *Journal of the Science of Food and Agriculture*, vol. 90, no. 15, 2010, pp. 2660–2663., doi:10.1002/jsfa.4136.
61. Kim, Ha Won, et al. "Gene Expressions of Taurine Transporter and Taurine Biosynthetic Enzyme during Mouse and Chicken Embryonic Development." *Taurine 6*, pp. 69–77., doi:10.1007/978-0-387-33504-9_7.
62. Dai, Bin, et al. "Influence of Dietary Taurine and Housing Density on Oviduct Function in Laying Hens." *Journal of Zhejiang University-SCIENCE B*, vol. 16, no. 6, 2015, pp. 456–464., doi:10.1631/jzus.b1400256.
63. Narvaez-So, William, et al. "Nutritional Requirements in Methionine + Cystine for White-Egg Laying Hens during the First Cycle of Production." *International Journal of Poultry Science*, vol. 4, no. 12, 2005, pp. 965–968., doi:10.3923/ijps.2005.965.968.
64. Harms, R.H., et al. "The Influence of Methionine on Commercial Laying Hens." *Journal of Applied Poultry Research*, vol. 7, no. 1, 1998, pp. 45–52., doi:10.1093/japr/7.1.45.
65. Fisher, James R., and Robert E. Eakin. "Nitrogen Excretion in Developing Chick Embryos." *Development*, vol. 5, no. 3, 1957, pp. 215–224., doi:10.1242/dev.5.3.215.
66. NEEDHAM, JOSEPH, et al. "The Origin and Fate of Urea in the Developing Hen's Egg." *Journal of Experimental Biology*, vol. 12, no. 4, 1935, pp. 321–336., doi:10.1242/jeb.12.4.321.
67. Sartori, M.R., et al. "Nitrogen Excretion during Embryonic Development of the Green Iguana, Iguana Iguana (Reptilia; Squamata)." *Comparative Biochemistry and Physiology Part A: Molecular & Integrative Physiology*, vol. 163, no. 2, 2012, pp. 210–214., doi:10.1016/j.cbpa.2012.06.004.
68. Clark, Hugh, et al. "Excretion of Nitrogen by the Alligator Embryo." *Journal of Cellular and Comparative Physiology*, vol. 50, no. 1, 1957, pp. 129–134., doi:10.1002/jcp.1030500109.
69. Ezzell, Carol. "The Himba and the Dam." *Scientific American*, vol. 284, no. 6, 2001, pp. 80–89., doi:10.1038/scientificamerican0601-80.

70. Woese, C. R., and G. E. Fox. "Phylogenetic Structure of the Prokaryotic Domain: The Primary Kingdoms." *Proceedings of the National Academy of Sciences*, vol. 74, no. 11, 1977, pp. 5088–5090., doi:10.1073/pnas.74.11.5088.
71. Woese, C. R., et al. "Towards a Natural System of Organisms: Proposal for the Domains Archaea, Bacteria, and Eucarya." *Proceedings of the National Academy of Sciences*, vol. 87, no. 12, 1990, pp. 4576–4579., doi:10.1073/pnas.87.12.4576.
72. De Duve, Christian. *Life Evolving: Molecules, Mind, and Meaning.* Oxford University Press, Inc., 2002, pp. 120-129.
73. Williams, Kelly P., et al. "Phylogeny of Gammaproteobacteria." *Journal of Bacteriology*, vol. 192, no. 9, 2010, pp. 2305–2314., doi:10.1128/jb.01480-09.
74. Chaikeeratisak, Vorrapon, et al. "Assembly of a Nucleus-like Structure during Viral Replication in Bacteria." *Science*, vol. 355, no. 6321, 2017, pp. 194–197., doi:10.1126/science.aal2130.
75. Mendoza, Senén D., et al. "A Bacteriophage Nucleus-like Compartment Shields DNA from CRISPR Nucleases." *Nature*, vol. 577, no. 7789, 2019, pp. 244–248., doi:10.1038/s41586-019-1786-y.
76. Kraemer, James A., et al. "A Phage Tubulin Assembles Dynamic Filaments by an Atypical Mechanism to Center Viral DNA within the Host Cell." *Cell*, vol. 149, no. 7, 2012, pp. 1488–1499., doi:10.1016/j.cell.2012.04.034.
77. Chaikeeratisak, Vorrapon, et al. "The Phage Nucleus and Tubulin Spindle Are Conserved among Large Pseudomonas Phages." *Cell Reports*, vol. 20, no. 7, 2017, pp. 1563–1571., doi:10.1016/j.celrep.2017.07.064.
78. Pilhofer, Martin, et al. "Microtubules in Bacteria: Ancient Tubulins Build a Five-Protofilament Homolog of the Eukaryotic Cytoskeleton." *PLoS Biology*, vol. 9, no. 12, 2011, doi:10.1371/journal.pbio.1001213.
79. Karnkowska, Anna, et al. "A Eukaryote without a Mitochondrial Organelle." *Current Biology*, vol. 26, no. 10, 2016, pp. 1274–1284., doi:10.1016/j.cub.2016.03.053.

80. Muñoz-Gómez, Sergio A., et al. "The Origin of Mitochondrial Cristae from Alphaproteobacteria." *Molecular Biology and Evolution*, 2017, doi:10.1093/molbev/msw298.
81. Thrash, J. Cameron, et al. "Phylogenomic Evidence for a Common Ancestor of Mitochondria and the SAR11 Clade." *Scientific Reports*, vol. 1, no. 1, 2011, doi:10.1038/srep00013.
82. Kang, I., et al. "Genome of a sar116 Bacteriophage Shows the Prevalence of This Phage Type in the Oceans." *Proceedings of the National Academy of Sciences*, vol. 110, no. 30, 2013, pp. 12343–12348., doi:10.1073/pnas.1219930110.
83. Zhao, Yanlin, et al. "Pelagiphages in ThePodoviridaefamily Integrate into Host Genomes." *Environmental Microbiology*, vol. 21, no. 6, 2018, pp. 1989–2001., doi:10.1111/1462-2920.14487.
84. Martin, William F., et al. "Endosymbiotic Theories for Eukaryote Origin." *Philosophical Transactions of the Royal Society B: Biological Sciences*, vol. 370, no. 1678, 2015, p. 20140330., doi:10.1098/rstb.2014.0330.
85. Zinder, Norton D., and Joshua Lederberg. "Genetic Exchange in Salmonella." *Journal of Bacteriology*, vol. 64, no. 5, 1952, pp. 679–699., doi:10.1128/jb.64.5.679-699.1952.
86. Miller, R.V. Environmental bacteriophage-host interactions: factors contribution to natural transduction. *Antonie Van Leeuwenhoek* 79, 141–147 (2001). https://doi.org/10.1023/A:1010278628468
87. MacGregor, Barbara J., et al. "Mobile Elements in a Single-Filament Orange Guaymas Basin Beggiatoa ('Candidatus Maribeggiatoa') Sp.. Draft Genome: Evidence for Genetic Exchange with Cyanobacteria." *Applied and Environmental Microbiology*, vol. 79, no. 13, 2013, pp. 3974–3985., doi:10.1128/aem.03821-12.
88. Chibani-Chennoufi, Sandra, et al. "Phage-Host Interaction: An Ecological Perspective." *Journal of Bacteriology*, vol. 186, no. 12, 2004, pp. 3677–3686., doi:10.1128/jb.186.12.3677-3686.2004.
89. Breitbart, Mya, et al. "Exploring the Vast Diversity of Marine Viruses." *Oceanography*, vol. 20, no. 2, 2007, pp. 135–139., doi:10.5670/oceanog.2007.58.

Chapter 2

1. Taylor, Martha R, et. al. Campbell Biology: Concepts & Connections. (9th Edition). *Pearson*. 2018, p. 334. Print.
2. Gehling, James G., and J. Keith Rigby. "Long Expected Sponges from the Neoproterozoic Ediacara Fauna of South Australia." *Journal of Paleontology*, vol. 70, no. 2, 1996, pp. 185–195., doi:10.1017/s0022336000023283.
3. Li, C. "Precambrian Sponges with Cellular Structures." *Science*, vol. 279, no. 5352, 1998, pp. 879–882., doi:10.1126/science.279.5352.879.
4. Bailey, Jake V., et al. "Evidence of Giant Sulphur Bacteria in Neoproterozoic Phosphorites." *Nature*, vol. 445, no. 7124, 2006, pp. 198–201., doi:10.1038/nature05457.
5. Knoll, Andrew H. *Life on a Young Planet: The First Three Billion Years of Evolution on Earth*. Princeton University Press, 2015, pp 147-148
6. Gold, David A., et al. "Sterol and Genomic Analyses Validate the Sponge Biomarker Hypothesis." *Proceedings of the National Academy of Sciences*, vol. 113, no. 10, 2016, pp. 2684–2689., doi:10.1073/pnas.1512614113.
7. Maldonado, Manuel, and Ana Riesgo. "*Reproduction in the phylum Porifera: A synoptic overview*." Treballs de la Societat Catalana de Biologia. Vol. 59, 2008, pp. 29-49, p. 45; doi: 10.2436/20.1501.02.56.
8. Ereskovsky, Alexander V. "Sponge Reproduction." *Encyclopedia of Reproduction*, 2018, pp. 485–490., doi:10.1016/b978-0-12-809633-8.20596-7.
9. Wilkin, Douglas, and Jennifer Blanchette. "Sponge Reproduction." *CK*, CK-12 Foundation, 20 Apr. 2020. https://ck12.org/biology/sponge-reproduction/lesson/Sponge-Reproduction-Advanced-BIO-ADV
10. Bailey, Jake V, et al. "Dimorphism in Methane Seep-Dwelling Ecotypes of the Largest Known Bacteria." *The ISME Journal*, vol. 5, no. 12, 2011, pp. 1926–1935., doi:10.1038/ismej.2011.66.
11. Arak, Anthony. "Male-Male Competition and Mate Choice in Anuran Amphibians." *Mate Choice*, by Patrick Bateson, Cambridge University Press, 1983, pp. 181–210.

12. Browne, R.K., et al. "Sperm Motility of Externally Fertilizing Fish and Amphibians." *Theriogenology*, vol. 83, no. 1, 2015, doi:10.1016/j.theriogenology.2014.09.018.
13. Barrière, Antoine, and Marie-Anne Félix. "High Local Genetic Diversity and Low Outcrossing Rate in Caenorhabditis Elegans Natural Populations." *Current Biology*, vol. 15, no. 13, 2005, pp. 1176–1184., doi:10.1016/j.cub.2005.06.022.
14. Sivasundar, Arjun, and Jody Hey. "Sampling from Natural Populations with Rnai Reveals High Outcrossing and Population Structure in Caenorhabditis Elegans." *Current Biology*, vol. 15, no. 17, 2005, pp. 1598–1602., doi:10.1016/j.cub.2005.08.034.
15. Bahrami, Adam K, and Yun Zhang. "When Females Produce Sperm: Genetics of C. Elegans Hermaphrodite Reproductive Choice." *G3 Genes|Genomes|Genetics*, vol. 3, no. 10, 2013, pp. 1851–1859., doi:10.1534/g3.113.007914.
16. Jarne, Philippe, and Josh R. Auld. "Animals Mix It up Too: The Distribution of Self-Fertilization among Hermaphroditic Animals." *Evolution*, vol. 60, no. 9, 2006, p. 1816., doi:10.1554/06-246.1.
17. NC State, Agriculture and Life Sciences. "Reproductive System." *ENT 425 - General Entomology*. https://genent.cals.ncsu.edu/bug-bytes/reproductive-system/. Accessed online 8/9/2020.
18. Clift, Dean, and Melina Schuh. "Restarting Life: Fertilization and the Transition from Meiosis to Mitosis." *Nature Reviews Molecular Cell Biology*, vol. 14, no. 9, 2013, pp. 549–562., doi:10.1038/nrm3643.
19. Blond, Jean-Luc, et al. "Molecular Characterization and Placental Expression of HERV-w, a New Human Endogenous Retrovirus Family." *Journal of Virology*, vol. 73, no. 2, 1999, pp. 1175–1185., doi:10.1128/jvi.73.2.1175-1185.1999.
20. Mi, Sha, et al. "Syncytin Is a Captive Retroviral Envelope Protein Involved in Human Placental Morphogenesis." *Nature*, vol. 403, no. 6771, 2000, pp. 785–789., doi:10.1038/35001608.
21. Vargas, Amandine, et al. "Syncytin-2 Plays an Important Role in the Fusion of Human Trophoblast Cells." *Journal of Molecular Biology*, vol. 392, no. 2, 2009, pp. 301–318., doi:10.1016/j.jmb.2009.07.025.

22. Okada, Y., and F. Murayama. "Multinucleated Giant Cell Formation by Fusion between Cells of Two Different Strains." *Experimental Cell Research*, vol. 40, no. 1, 1965, pp. 154–158., doi:10.1016/0014-4827(65)90303-4.
23. Iowa State University. *Sendai Virus*. 2015. https://swinehealth.org/wp-content/uploads/2016/03/Sendai-virus-SeV.pdf. Accessed online 9/5/2021
24. Hallett, Maurice B., et al. "Sendai Virus Causes a Rise in Intracellular Free ca2+ before Cell Fusion." *Biochemical Journal*, vol. 206, no. 3, 1982, pp. 671–674., doi:10.1042/bj2060671.
25. Shahrabadi, M S, and P W Lee. "Calcium Requirement for Syncytium Formation in HEP-2 Cells by Respiratory Syncytial Virus." *Journal of Clinical Microbiology*, vol. 26, no. 1, 1988, pp. 139–141., doi:10.1128/jcm.26.1.139-141.1988.
26. Dimitrov, D S, et al. "Calcium Ions Are Required for Cell Fusion Mediated by the CD4-Human Immunodeficiency Virus Type 1 Envelope Glycoprotein Interaction." *Journal of Virology*, vol. 67, no. 3, 1993, pp. 1647–1652., doi:10.1128/jvi.67.3.1647-1652.1993.
27. Kutter, Elizabeth, and Alexander Sulakvelidze. *Bacteriophages: Biology and Applications*. CRC Press, 2005.
28. Finnerty, J. R. "Origins of Bilateral Symmetry: Hox and DPP Expression in a Sea Anemone." *Science*, vol. 304, no. 5675, 2004, pp. 1335–1337., doi:10.1126/science.1091946.
29. *Introduction to the Bilateria and the Phylum Xenacoelomorpha* . Sinauer Associates, Inc., 2016. https://www.sinauer.com/media/wysiwyg/samples/Brusca3e_Chapter_9.pdf. Accessed online on 3/3/2021
30. Padgett, Richard W., et al. "A Transcript from a Drosophila Pattern Gene Predicts a Protein Homologous to the Transforming Growth Factor-β Family." *Nature*, vol. 325, no. 6099, 1987, pp. 81–84., doi:10.1038/325081a0.
31. Irish, V F, and W M Gelbart. "The Decapentaplegic Gene Is Required for Dorsal-Ventral Patterning of the Drosophila Embryo." *Genes & Development*, vol. 1, no. 8, 1987, pp. 868–879., doi:10.1101/gad.1.8.868.

32. Ryan, Joseph F., et al. "Pre-Bilaterian Origins of the Hox Cluster and the Hox Code: Evidence from the Sea Anemone, Nematostella Vectensis." *PLoS ONE*, vol. 2, no. 1, 2007, doi:10.1371/journal.pone.0000153.
33. Rentzsch, Fabian, and Thomas W. Holstein. "Making Head or Tail of Cnidarian Hox Gene Function." *Nature Communications*, vol. 9, no. 1, 2018, doi:10.1038/s41467-018-04585-y.
34. Ackermann, Hans-W. "Classification of Bacteriophages." *The Bacteriophages*, by Richard Calendar, 2nd ed., Oxford University Press, 2006, p. 9.

Chapter 3

1. Fujito, Satoshi, et al. "Evidence for a Common Origin of Homomorphic and Heteromorphic Sex Chromosomes in Distinct Spinacia Species." *G3 Genes|Genomes|Genetics*, vol. 5, no. 8, 2015, pp. 1663–1673., doi:10.1534/g3.115.018671.
2. Hodgkin, Jonathan. "Sex Determination in the Nematode C. Elegans: Analysis of TRA-3 Suppressors and Characterization of FEM Genes." *Genetics*, vol. 114, no. 1, 1986, pp. 15–52., doi:10.1093/genetics/114.1.15.
3. Bahrami, Adam K, and Yun Zhang. "When Females Produce Sperm: Genetics of C. Elegans Hermaphrodite Reproductive Choice." *G3 Genes|Genomes|Genetics*, vol. 3, no. 10, 2013, pp. 1851–1859., doi:10.1534/g3.113.007914.
4. Jarne, Philippe, and Josh R. Auld. "Animals Mix It up Too: The Distribution of Self-Fertilization among Hermaphroditic Animals." *Evolution*, vol. 60, no. 9, 2006, p. 1816., doi:10.1554/06-246.1.
5. Saccheri, Ilik, et al. "Inbreeding and Extinction in a Butterfly Metapopulation." *Nature*, vol. 392, no. 6675, 1998, pp. 491–494., doi:10.1038/33136.
6. Nabulsi, Mona M., et al. "Parental Consanguinity and Congenital Heart Malformations in a Developing Country." *American Journal of Medical Genetics Part A*, vol. 116A, no. 4, 2001, pp. 342–347., doi:10.1002/ajmg.a.10020.

7. Jimenez, J., et al. "An Experimental Study of Inbreeding Depression in a Natural Habitat." *Science*, vol. 266, no. 5183, 1994, pp. 271–273., doi:10.1126/science.7939661.
8. Chen, Xiaofeng. "Comparison of Inbreeding and Outbreeding in Hermaphroditic Arianta Arbustorum (L.) (Land Snail)." *Heredity*, vol. 71, no. 5, 1993, pp. 456–461., doi:10.1038/hdy.1993.163.
9. "Porifera: Life History and Ecology." *Life History and Ecology of Porifera*. https://ucmp.berkeley.edu/porifera/poriferalh.html. Accessed online 8/13/2020.
10. Maldonado, Manuel, and Ana Riesgo. "*Reproduction in the phylum Porifera: A synoptic overview.*" Treballs de la Societat Catalana de Biologia. Vol. 59, 2008, pp. 29-49, p. 43; doi: 10.2436/20.1501.02.56.
11. Renner, Susanne S., and Robert E. Ricklefs. "Dioecy and Its Correlates in the Flowering Plants." *American Journal of Botany*, vol. 82, no. 5, 1995, pp. 596–606., doi:10.1002/j.1537-2197.1995.tb11504.x.
12. Pannell, John R. "Plant Sex Determination." *Current Biology*, vol. 27, no. 5, 2017, doi:10.1016/j.cub.2017.01.052.
13. Bachtrog, Doris. "Y-Chromosome Evolution: Emerging Insights into Processes of Y-Chromosome Degeneration." *Nature Reviews Genetics*, vol. 14, no. 2, 2013, pp. 113–124., doi:10.1038/nrg3366.
14. Hasselmann, Martin, et al. "Evidence for the Evolutionary Nascence of a Novel Sex Determination Pathway in Honeybees." *Nature*, vol. 454, no. 7203, 2008, pp. 519–522., doi:10.1038/nature07052.
15. Schmieder, Sandra, et al. "Tracing Back the Nascence of a New Sex-Determination Pathway to the Ancestor of Bees and Ants." *Nature Communications*, vol. 3, no. 1, 2012, doi:10.1038/ncomms1898.
16. Virant-Klun, Irma. "Postnatal Oogenesis in Humans: A Review of Recent Findings." *Stem Cells and Cloning: Advances and Applications*, 2015, p. 49., doi:10.2147/sccaa.s32650.
17. Clift, Dean, and Melina Schuh. "Restarting Life: Fertilization and the Transition from Meiosis to Mitosis." *Nature Reviews Molecular Cell Biology*, vol. 14, no. 9, 2013, pp. 549–562., doi:10.1038/nrm3643.
18. MM;, Perry. "Nuclear Events from Fertilisation to the Early Cleavage Stages in the Domestic Fowl (Gallus Domesticus)." *Journal of*

Anatomy, U.S. National Library of Medicine. https://pubmed.ncbi.nlm.nih.gov/2443474/

19. Nakamura, Yoshiaki, et al. "Development, Differentiation and Manipulation of Chicken Germ Cells." *Development, Growth & Differentiation*, vol. 55, no. 1, 2013, pp. 20–40., doi:10.1111/dgd.12026.
20. VERMA, Savitri, and Friedrich RUTTNER. "Cytological Analysis of the Thelytokous Parthenogenesis in the Cape Honeybee (Apis Mellifera Capensis ESCHOLTZ)." *Apidologie*, vol. 14, no. 1, 1983, pp. 41–57., doi:10.1051/apido:19830104.
21. Snodgrass, Robert Evans. *The Anatomy of the Honey Bee.* May 1910. https://naldc.nal.usda.gov/download/CAT31027153/PDF
22. Oldroyd, Benjamin P., et al. "Viable Triploid Honey Bees (Apis Mellifera Capensis) Are Reliably Produced in the Progeny of CO2 Narcotised Queens." *G3*, G3: Genes, Genomes, Genetics, 1 Oct. 2018. https://g3journal.org/content/8/10/3357
23. Goudie, Frances, et al. "Maintenance and Loss of Heterozygosity in a Thelytokous Lineage of Honey Bees (Apis Mellifera Capensis)." *Wiley Online Library*, John Wiley & Sons, Ltd, 2 Feb. 2012. https://onlinelibrary.wiley.com/doi/full/10.1111/j.1558-5646.2011.01543.x
24. Cole-Clark, Miles P., et al. "Cytogenetic Basis of Thelytoky in Apis Mellifera Capensis." *Apidologie*, vol. 48, no. 5, 2017, pp. 623–634., doi:10.1007/s13592-017-0505-7.
25. "Cornu Aspersum." *Wikipedia*. Wikimedia Foundation. https://en.wikipedia.org/wiki/Cornu_aspersum. Accessed online on 10/2/2019.
26. Otto, Sarah P. "The Evolutionary Enigma of Sex." *The American Naturalist*, vol. 174, no. S1, 2009, doi:10.1086/599084.
27. Chue, Justin, and Craig A. Smith. "Sex Determination and Sexual Differentiation in the Avian Model." *FEBS Journal*, vol. 278, no. 7, 2011, pp. 1027–1034., doi:10.1111/j.1742-4658.2011.08032.x.
28. Fridolfsson, A.-K., et al. "Evolution of the Avian Sex Chromosomes from an Ancestral Pair of Autosomes." *Proceedings of the National Academy of Sciences*, vol. 95, no. 14, 1998, pp. 8147–8152., doi:10.1073/pnas.95.14.8147.

29. Graves, J A. "The Evolution of Mammalian Sex Chromosomes and the Origin of Sex Determining Genes." *Philosophical Transactions of the Royal Society of London. Series B: Biological Sciences*, vol. 350, no. 1333, 1995, pp. 305–312., doi:10.1098/rstb.1995.0166.
30. Abbott, Jessica K., et al. "Sex Chromosome Evolution: Historical Insights and Future Perspectives." *Proceedings of the Royal Society B: Biological Sciences*, vol. 284, no. 1854, 2017, p. 20162806., doi:10.1098/rspb.2016.2806.
31. Kaiser, Vera B., and Doris Bachtrog. "Evolution of Sex Chromosomes in Insects." *Annual Review of Genetics*, vol. 44, no. 1, 2010, pp. 91–112., doi:10.1146/annurev-genet-102209-163600.
32. Hodgkin, Jonathan A, and Sydney Brenner. "Mutations Causing Transformation of Sexual Phenotype in the Nematode Caenorhabditis Elegans." *Genetics*, vol. 86, no. 2, 1977, pp. 275–287., doi:10.1093/genetics/86.2.275.
33. Pomiankowski, A. "The Evolution of the Drosophila Sex-Determination Pathway." *Genetics*, vol. 166, no. 4, 2004, pp. 1761–1773., doi:10.1534/genetics.166.4.1761.
34. Gilbert, Scott F. "Chromosomal Sex Determination in Drosophila." *Developmental Biology. 6th Edition.*, U.S. National Library of Medicine, 1 Jan. 1970. https://ncbi.nlm.nih.gov/books/NBK10025/
35. Erickson, James W, and Jerome J Quintero. "Indirect Effects of Ploidy Suggest X Chromosome Dose, Not the x:a Ratio, Signals Sex in Drosophila." *PLoS Biology*, vol. 5, no. 12, 2007, doi:10.1371/journal.pbio.0050332.
36. Cline, Thomas W. "The Drosophila Sex Determination Signal: How Do Flies Count to Two?" *Trends in Genetics*, vol. 9, no. 11, 1993, pp. 385–390., doi:10.1016/0168-9525(93)90138-8.
37. Beye, Martin, et al. "The Gene CSD Is the Primary Signal for Sexual Development in the Honeybee and Encodes an SR-Type Protein." *Cell*, vol. 114, no. 4, 2003, pp. 419–429., doi:10.1016/s0092-8674(03)00606-8.
38. Hasselmann, Martin, and Martin Beye. "Pronounced Differences of Recombination Activity at the Sex Determination Locus of the

Honeybee, a Locus under Strong Balancing Selection." *Genetics*, vol. 174, no. 3, 2006, pp. 1469–1480., doi:10.1534/genetics.106.062018.

39. KATSUMA, Susumu, et al. "Unique Sex Determination System in the Silkworm, *Bombyx Mori*: Current Status and Beyond." *Proceedings of the Japan Academy, Series B*, vol. 94, no. 5, 2018, pp. 205–216., doi:10.2183/pjab.94.014.

Chapter 4

1. Eldredge, N., and S. J. Gould. "Punctuated Equilibria: an Alternative to Phyletic Gradualism." *Models in Paleobiology*, by Schopf Thomas J M., Freeman, Cooper, 1972, p. 93.
2. Taylor, Martha R, et. al. Campbell Biology: Concepts & Connections. (9th Edition). *Pearson*. 2018, p. 283. Print.
3. Mora, Camilo, et al. "How Many Species Are There on Earth and in the Ocean?" *PLoS Biology*, vol. 9, no. 8, 2011, doi:10.1371/journal.pbio.1001127.
4. Quackenbush, John. *The Human Genome: The Book of Essential Knowledge.* Imagine, 2011, pp. 66-67
5. The Chimpanzee Sequencing and Analysis Consortium. Initial sequence of the chimpanzee genome and comparison with the human genome. *Nature* 437, 69–87 (2005). https://doi.org/10.1038/nature04072
6. Darwin, Charles. *On the Origin of Species by Means of Natural Selection, or Preservation of Favoured Races in the Struggle for Life.* John Murray, Albemarle Street, 1859, p. 109. http://darwin-online.org.uk/converted/pdf/1859_Origin_F373.pdf
7. Voet, Donald, et al. *Fundamentals of Biochemistry: Life at the Molecular Level.* John Wiley & Sons, 2016, p. 909
8. Drake, John W, et al. "Rates of Spontaneous Mutation." *Genetics*, vol. 148, no. 4, 1998, p. 1674., doi:10.1093/genetics/148.4.1667.
9. Muller, H. J. "Artificial Transmutation of the Gene." *Science*, vol. 66, no. 1699, 1927, pp. 84–87., doi:10.1126/science.66.1699.84.
10. Dobzhansky, Theodosius Grigorievich. *Evolution, Genetics and Man.* Wiley, 1965, p. 105.

11. Coyne, Jerry A. *Why Evolution Is True.* Oxford University Press, 2010, p. 217.
12. Swallow, Dallas M. "Genetics of Lactase Persistence and Lactose Intolerance." *Annual Review of Genetics*, vol. 37, no. 1, 2003, pp. 197–219., doi:10.1146/annurev.genet.37.110801.143820.
13. Origins of Lactose Intolerance. https://chesteruplandsd.org/userfiles/-17/my%20files/origins%20of%20lactose%20intolerance.pdf?id=879. Accessed online 8/2020.
14. Cross, Steve, et al. *Lactose Tolerance.* https://dnadarwin.org/casestudies/5/FILES/LactaseTG1.1.pdf. Accessed online 8/2020.
15. Gerbault, Pascale, et al. "Evolution of Lactase Persistence: An Example of Human Niche Construction." *Philosophical Transactions of the Royal Society B: Biological Sciences*, vol. 366, no. 1566, 2011, pp. 863–877., doi:10.1098/rstb.2010.0268.
16. Mortlock, Robert P. *Microorganisms as Model Systems for Studying Evolution*. Plenum Press, 1984.
17. Fossitt, Dexter, et al. "Pathways of L-Arabitol and Xylitol Metabolism in Aerobacter Aerogenes." *Journal of Biological Chemistry*, vol. 239, no. 7, 1964, pp. 2110–2115., doi:10.1016/s0021-9258(20)82207-8.
18. Mortlock, R. P., et al. "A Basis for Utlization of Unnatural Pentoses and Pentitols by Aerobacter Aerogenes." *Proceedings of the National Academy of Sciences*, vol. 54, no. 2, 1965, pp. 572–579., doi:10.1073/pnas.54.2.572.
19. Wu, T T, et al. "Mutants of Aerobacter Aerogenes Capable of Utilizing Xylitol as a Novel Carbon." *Journal of Bacteriology*, U.S. National Library of Medicine, Aug. 1968. https://ncbi.nlm.nih.gov/pubmed/5674056
20. Platt, Orah S., et al. "Mortality in Sickle Cell Disease — Life Expectancy and Risk Factors for Early Death." *New England Journal of Medicine*, vol. 330, no. 23, 1994, pp. 1639–1644., doi:10.1056/nejm199406093302303.
21. Lobo, Clarisse Lopes, et al. "Mortality in Children, Adolescents and Adults with Sickle Cell Anemia in Rio De Janeiro, Brazil." *Hematology, Transfusion and Cell Therapy*, vol. 40, no. 1, 2018, pp. 37–42., doi:10.1016/j.bjhh.2017.09.006.

22. Luzzatto, Lucio. "Sickle Cell Anaemia and Malaria." *Mediterranean Journal of Hematology and Infectious Diseases*, Università Cattolica Del Sacro Cuore, 2012. https://ncbi.nlm.nih.gov/pmc/articles/PMC3499995/
23. Friedman, M. J. "Erythrocytic Mechanism of Sickle Cell Resistance to Malaria." *Proceedings of the National Academy of Sciences*, vol. 75, no. 4, 1978, pp. 1994–1997., doi:10.1073/pnas.75.4.1994.
24. Lauring, Adam S., and Raul Andino. "Quasispecies Theory and the Behavior of RNA Viruses." *PLoS Pathogens*, vol. 6, no. 7, 2010, doi:10.1371/journal.ppat.1001005.
25. AKINYANJU, OLUFEMI O. "A Profile of Sickle Cell Disease in Nigeria." *Annals of the New York Academy of Sciences*, vol. 565, no. 1 Sickle Cell D, 1989, pp. 126–136., doi:10.1111/j.1749-6632.1989.tb24159.x.
26. Farrell, Philip M. "The Prevalence of Cystic Fibrosis in the European Union." *Journal of Cystic Fibrosis*, vol. 7, no. 5, 2008, pp. 450–453., doi:10.1016/j.jcf.2008.03.007.
27. Oteng-Ntim, E., et al. "Adverse Maternal and Perinatal Outcomes in Pregnant Women with Sickle Cell Disease: Systematic Review and Meta-Analysis." *Blood*, vol. 125, no. 21, 2015, pp. 3316–3325., doi:10.1182/blood-2014-11-607317.
28. Fisher, Amelia E., et al. "Presentations of Sickle Cell Disease Patients to Hospital in Ghana: Key Findings from a Preliminary Study at Volta Regional Hospital." *British Journal of Haematology*, vol. 178, no. 3, 2016, pp. 489–491., doi:10.1111/bjh.14154.
29. Silva, Flávia A.C., et al. "Adverse Clinical and Obstetric Outcomes among Pregnant Women with Different Sickle Cell Disease Genotypes." *International Journal of Gynecology & Obstetrics*, vol. 143, no. 1, 2018, pp. 89–93., doi:10.1002/ijgo.12626.
30. Silva-Pinto, Ana Cristina, et al. "Sickle Cell Disease and Pregnancy: Analysis of 34 Patients Followed at the Regional Blood Center of Ribeirão Preto, Brazil." *Revista Brasileira De Hematologia e Hemoterapia*, vol. 36, no. 5, 2014, pp. 329–333., doi:10.1016/j.bjhh.2014.07.002.

31. Jordan, D. S. "The Origin of Species through Isolation." *Science*, vol. 22, no. 566, 1905, pp. 545–562., doi:10.1126/science.22.566.545.
32. Lamichhaney, Sangeet, et al. "Evolution of Darwin's Finches and Their Beaks Revealed by Genome Sequencing." *Nature*, vol. 518, no. 7539, 2015, pp. 371–375., doi:10.1038/nature14181.
33. Malinsky, Milan, et al. "Whole-Genome Sequences of Malawi Cichlids Reveal Multiple Radiations Interconnected by Gene Flow." *Nature Ecology & Evolution*, vol. 2, no. 12, 2018, pp. 1940–1955., doi:10.1038/s41559-018-0717-x.
34. Genner, M. J., and G. F. Turner. "Ancient Hybridization and Phenotypic Novelty within Lake Malawi's Cichlid Fish Radiation." *Molecular Biology and Evolution*, vol. 29, no. 1, 2011, pp. 195–206., doi:10.1093/molbev/msr183.
35. Schilthuizen, Menno. *Frogs, Flies and Dandelions: Speciation - the Evolution of New Species*. Oxford University Press, 2006.
36. Sternkopf, Viviane, et al. "Introgressive Hybridization and the Evolutionary History of the Herring Gull Complex Revealed by Mitochondrial and Nuclear DNA." *BMC Evolutionary Biology*, vol. 10, no. 1, 2010, doi:10.1186/1471-2148-10-348.
37. Sonsthagen, Sarah A., et al. "Hybridization among Arctic White-Headed Gulls (Larusspp.) Obscures the Genetic Legacy of the Pleistocene." *Ecology and Evolution*, vol. 2, no. 6, 2012, pp. 1278–1295., doi:10.1002/ece3.240.
38. Darwin, Charles. *The Voyage of the Beagle*. P.F. Collier & Son, 1909.
39. Bebenek, K, et al. "Specificity and Mechanism of Error-Prone Replication by Human Immunodeficiency Virus-1 Reverse Transcriptase." *Journal of Biological Chemistry*, vol. 264, no. 28, 1989, pp. 16948–16956., doi:10.1016/s0021-9258(19)84799-3.
40. Lovelock, J.E. "Gaia as Seen through the Atmosphere." *Atmospheric Environment (1967)*, vol. 6, no. 8, 1972, pp. 579–580., doi:10.1016/0004-6981(72)90076-5.
41. Freeland, Stephen J., and Laurence D. Hurst. "The Genetic Code Is One in a Million." *Journal of Molecular Evolution*, vol. 47, no. 3, 1998, pp. 238–248., doi:10.1007/pl00006381.

42. Holland, J, et al. “Rapid Evolution of RNA Genomes.” *Science*, vol. 215, no. 4540, 1982, pp. 1577–1585., doi:10.1126/science.7041255.
43. Poch, O., et al. “Identification of Four Conserved Motifs among the RNA-Dependent Polymerase Encoding Elements.” *The EMBO Journal*, vol. 8, no. 12, 1989, pp. 3867–3874., doi:10.1002/j.1460-2075.1989.tb08565.x.
44. Duffy, Siobain, et al. “Rates of Evolutionary Change in Viruses: Patterns and Determinants.” *Nature Reviews Genetics*, vol. 9, no. 4, 2008, pp. 267–276., doi:10.1038/nrg2323.
45. Domingo, E., et al. “Viruses as Quasispecies: Biological Implications.” *Current Topics in Microbiology and Immunology*, 2006, pp. 51–82., doi:10.1007/3-540-26397-7_3.
46. Vignuzzi, Marco, et al. “Quasispecies Diversity Determines Pathogenesis through Cooperative Interactions in a Viral Population.” *Nature*, vol. 439, no. 7074, 2005, pp. 344–348., doi:10.1038/nature04388.
47. Webster , R G, et al. “Evolution and Ecology of Influenza A Viruses.” *Microbiological Reviews*, U.S. National Library of Medicine. https://pubmed.ncbi.nlm.nih.gov/1579108/
48. Chen, Rubing, and Edward C. Holmes. “Avian Influenza Virus Exhibits Rapid Evolutionary Dynamics.” *Molecular Biology and Evolution*, vol. 23, no. 12, 2006, pp. 2336–2341., doi:10.1093/molbev/msl102.
49. Murphy, B. R., and M. L. Clements. “The Systemic and Mucosal Immune Response of Humans to Influenza A Virus.” *Current Topics in Microbiology and Immunology*, 1989, pp. 107–116., doi:10.1007/978-3-642-74529-4_12.
50. Taubenberger, Jeffery K., and John C. Kash. “Influenza Virus Evolution, Host Adaptation, and Pandemic Formation.” *Cell Host & Microbe*, vol. 7, no. 6, 2010, pp. 440–451., doi:10.1016/j.chom.2010.05.009.
51. Roberts, J., et al. “The Accuracy of Reverse Transcriptase from HIV-1.” *Science*, vol. 242, no. 4882, 1988, pp. 1171–1173., doi:10.1126/science.2460925.

52. Perelson, A. S., et al. "HIV-1 Dynamics in Vivo: Virion Clearance Rate, Infected Cell Life-Span, and Viral Generation Time." *Science*, vol. 271, no. 5255, 1996, pp. 1582–1586., doi:10.1126/science.271.5255.1582.
53. Murray, J. M., et al. "Timing of the Components of the HIV Life Cycle in Productively Infected CD4+ T Cells in a Population of HIV-Infected Individuals." *Journal of Virology*, vol. 85, no. 20, 2011, pp. 10798–10805., doi:10.1128/jvi.05095-11.
54. Eigen, Manfred. "Viral Quasispecies." *Scientific American*, vol. 269, no. 1, 1993, pp. 42–49., doi:10.1038/scientificamerican0793-42.
55. Phillips, David M, and Gil L Dryden. "Comparative Morphology Of Mammalian Gametes." *A Comparative Overview of Mammalian Fertilization*, by Bonnie S. Dunbar and Michael G. O'Rand, Springer Science+Business Media, 2017, p. 37.
56. Chen, Jun-Yuan, et al. "Precambrian Animal Diversity: Putative Phosphatized Embryos from the Doushantuo Formation of China." *Proceedings of the National Academy of Sciences*, vol. 97, no. 9, 2000, pp. 4457–4462., doi:10.1073/pnas.97.9.4457.
57. Bailey, Jake V, et al. "Dimorphism in Methane Seep-Dwelling Ecotypes of the Largest Known Bacteria." *The ISME Journal*, vol. 5, no. 12, 2011, pp. 1926–1935., doi:10.1038/ismej.2011.66.
58. Bailey, Jake V., et al. "Evidence of Giant Sulphur Bacteria in Neoproterozoic Phosphorites." *Nature*, vol. 445, no. 7124, 2006, pp. 198–201., doi:10.1038/nature05457.

Chapter 5

1. Knoll, Andrew H. *Life on a Young Planet: The First Three Billion Years of Evolution on Earth*. Princeton University Press, 2015, p. 60-61
2. Javaux, Emmanuelle J., et al. "Morphological and Ecological Complexity in Early Eukaryotic Ecosystems." *Nature*, vol. 412, no. 6842, 2001, pp. 66–69., doi:10.1038/35083562.
3. PENG, Y, et al. "New Morphological Observations for Paleoproterozoic Acritarchs from the Chuanlinggou Formation, North China." *Precambrian Research*, vol. 168, no. 3-4, 2009, pp. 223–232., doi:10.1016/j.precamres.2008.10.005.

4. Lamb, D.M., et al. "Evidence for Eukaryotic Diversification in the ~1800 Million-Year-Old Changzhougou Formation, North China." *Precambrian Research*, vol. 173, no. 1-4, 2009, pp. 93–104., doi:10.1016/j.precamres.2009.05.005.
5. Javaux, Emmanuelle J., et al. "Organic-Walled Microfossils in 3.2-Billion-Year-Old Shallow-Marine Siliciclastic Deposits." *Nature*, vol. 463, no. 7283, 2010, pp. 934–938., doi:10.1038/nature08793.
6. Buick, Roger. "Ancient Acritarchs." *Nature*, vol. 463, no. 7283, 2010, pp. 885–886., doi:10.1038/463885a.
7. Evitt, W. R. "A Discussion and Proposals Concerning Fossil Dinoflagellates, Hystrichospheres, and Acritarchs, II." *Proceedings of the National Academy of Sciences*, vol. 49, no. 3, 1963, pp. 298–302., doi:10.1073/pnas.49.3.298.
8. Prothero, Donald R., and Robert H. Dott. *Evolution of the Earth.* McGraw-Hill, 2010, p. 191
9. Coyne, Jerry A. *Why Evolution Is True.* Oxford University Press, 2010, p. 183
10. "Molecular Cloud." *Wikipedia*, Wikimedia Foundation. https://en.wikipedia.org/wiki/Molecular_cloud#Occurrence. Accessed 1/19/21
11. Dobzhansky, Theodosius. *Genetics and the Origin of Species.* Columbia University Press, 1982, p. 11
12. Carrolls, Lewis. "Through the Looking-Glass." *The Project Gutenberg EBook of Through the Looking-Glass, by Lewis Carroll*, Chapter 2. https://www.gutenberg.org/files/12/12-h/12-h.htm#link2HCH0002
13. Taylor, Gordon Rattray. *The Great Evolution Mystery.* Harper & Row, 1983, pp. 115-119.
14. Prothero, Donald R., and Robert H. Dott. *Evolution of the Earth.* McGraw-Hill, 2010, pp. 57-58
15. Papagianni, Dimitra, and Michael A. Morse. *The Neanderthals Rediscovered: How Modern Science Is Rewriting Their Story.* Thames & Hudson, 2013, pp. 77-79.
16. Huxley, T. H. "On the Geographical Distribution of the Chief Modifications of Mankind." *The Journal of the Ethnological Society of London (1869-1870)*, vol. 2, no. 4, 1870, p. 404., doi:10.2307/3014371.

17. Zhuang , Ziqing, et al. "Facial Anthropometric Differences among Gender, Ethnicity, and Age Groups." *The Annals of Occupational Hygiene*, U.S. National Library of Medicine. https://pubmed.ncbi.nlm.nih.gov/20219836/
18. Coyne, Jerry A. *Why Evolution Is True*. Oxford University Press, 2010, p. 171.
19. Chen, Lu, et al. "Identifying and Interpreting Apparent Neanderthal Ancestry in African Individuals." *Cell*, vol. 180, no. 4, 2020, doi:10.1016/j.cell.2020.01.012.
20. Eldredge, Niles, and Stephen Jay Gould. "Punctuated Equilibria: an Alternative to Phyletic Gradualism." *Models in Paleobiology*, by Schopf Thomas J M., Freeman, Cooper, 1972.
21. Gould, Stephen Jay, and Niles Eldredge. "Punctuated Equilibria: The Tempo and Mode of Evolution Reconsidered." *Paleobiology*, vol. 3, no. 2, 1977, pp. 115–151., doi:10.1017/s0094837300005224.
22. Burney, D, and T Flannery. "Fifty Millennia of Catastrophic Extinctions after Human Contact." *Trends in Ecology & Evolution*, vol. 20, no. 7, 2005, pp. 395–401., doi:10.1016/j.tree.2005.04.022.
23. Demay, Laëtitia, et al. "Mammoths Used as Food and Building Resources by Neanderthals: Zooarchaeological Study Applied to Layer 4, Molodova I (Ukraine)." *Quaternary International*, vol. 276-277, 2012, pp. 212–226., doi:10.1016/j.quaint.2011.11.019.
24. Godefroit, P., et al. "A Jurassic Ornithischian Dinosaur from Siberia with Both Feathers and Scales." *Science*, vol. 345, no. 6195, 2014, pp. 451–455., doi:10.1126/science.1253351.
25. O'Connor, Patrick M., and Leon P. Claessens. "Basic Avian Pulmonary Design and Flow-through Ventilation in Non-Avian Theropod Dinosaurs." *Nature*, vol. 436, no. 7048, 2005, pp. 253–256., doi:10.1038/nature03716.
26. Meyer, Stephen C. *Darwin's Doubt: The Explosive Origin of Animal Life and the Case for Intelligent Design*. HarperOne, 2014, p. 311.
27. *Embryo-Fetal Circulation System - Changes at Birth. https*://embryology.ch/anglais/pcardio/umstellung02.html. Accessed online on 09/16/2020.

28. Jabr, Ferris. “How Does a Caterpillar Turn into a Butterfly?” *Scientific American*, Scientific American, 10 Aug. 2012. https://scientificamerican.com/article/caterpillar-butterfly-metamorphosis-explainer/
29. *Butterfly Life Cycle. https:*//ansp.org/exhibits/online-exhibits/butterflies/lifecycle/. Accessed online on 1/25/21
30. Darwin, Charles. *The Voyage of the Beagle.* P.F. Collier & Son, 1909, p. 178.
31. Darwin, Charles. *The Voyage of the Beagle.* P.F. Collier & Son, 1909, p. 180.
32. Zhou, Chang-Fu, et al. “A Jurassic Mammaliaform and the Earliest Mammalian Evolutionary Adaptations.” *Nature*, vol. 500, no. 7461, 2013, pp. 163–167., doi:10.1038/nature12429.
33. Luo, Zhe-Xi, et al. “Mandibular and Dental Characteristics of Late Triassic Mammaliaform Haramiyavia and Their Ramifications for Basal Mammal Evolution.” *Proceedings of the National Academy of Sciences*, vol. 112, no. 51, 2015, doi:10.1073/pnas.1519387112.
34. Payne, Rebecca C., et al. “Oxidized Micrometeorites Suggest Either High PCO2 or Low PN2 during the Neoarchean.” *Proceedings of the National Academy of Sciences*, vol. 117, no. 3, 2020, pp. 1360–1366., doi:10.1073/pnas.1910698117.
35. “Earth Science.” *Lumen.* https://courses.lumenlearning.com/earthscience/chapter/early-earth/. Accessed online on 9/24/2021.
36. Soo, Rochelle M., et al. “On the Origins of Oxygenic Photosynthesis and Aerobic Respiration in Cyanobacteria.” *Science*, vol. 355, no. 6332, 2017, pp. 1436–1440., doi:10.1126/science.aal3794.
37. Allen, Michael J, and William H Wilson. “Viruses of Algae and Mimivirus.” *Fundamentals of Molecular Virology*, by Nicholas H. Acheson, Wiley, 2011, pp. 331–332.
38. Mogire, Reagan M., et al. “Prevalence of Vitamin D Deficiency in Africa - A Systematic Review and Meta-Analysis.” *SSRN Electronic Journal*, 2019, doi:10.2139/ssrn.3382391.
39. Harris, Susan S. “Vitamin D and African Americans.” *The Journal of Nutrition*, vol. 136, no. 4, 2006, pp. 1126–1129., doi:10.1093/jn/136.4.1126.

40. Gordon, Louisa G., et al. "Modelling the Healthcare Costs of Skin Cancer in South Africa." *BMC Health Services Research*, vol. 16, no. 1, 2016, doi:10.1186/s12913-016-1364-z.
41. Norval, Mary, et al. "The Incidence and Body Site of Skin Cancers in the Population Groups of South Africa." *Photodermatology, Photoimmunology & Photomedicine*, vol. 30, no. 5, 2014, pp. 262–265., doi:10.1111/phpp.12106.
42. Li, Zi-Wen, et al. "Transposable Elements Contribute to the Adaptation of Arabidopsis Thaliana." *Genome Biology and Evolution*, vol. 10, no. 8, 2018, pp. 2140–2150., doi:10.1093/gbe/evy171.
43. Cornelis, Guy, et al. "TN951: A New Transposon Carrying a Lactose Operon." *Molecular and General Genetics MGG*, vol. 160, no. 2, 1978, pp. 215–224., doi:10.1007/bf00267484.
44. Saunders, J. R. "New Transposons." *Nature*, vol. 274, no. 5668, 1978, pp. 211–211., doi:10.1038/274211a0.
45. Feiss, Michael. "Bacteriophage Lambda." *Fundamentals of Molecular Virology*, by Nicholas H. Acheson, Wiley, 2011, p. 90.
46. Krishnan, Parvathy, et al. "Transposable Element Insertions Shape Gene Regulation and Melanin Production in a Fungal Pathogen of Wheat." *BMC Biology*, vol. 16, no. 1, 2018, doi:10.1186/s12915-018-0543-2.
47. Hof, Arjen E., et al. "The Industrial Melanism Mutation in British Peppered Moths Is a Transposable Element." *Nature*, vol. 534, no. 7605, 2016, pp. 102–105., doi:10.1038/nature17951.
48. "Kettlewell's Experiment." *Wikipedia*, Wikimedia Foundation. https://en.wikipedia.org/wiki/Kettlewell%27s_experiment#Criticisms. Accessed online on 9/14/2020
49. "Peppered Moth Evolution." *Wikipedia*, Wikimedia Foundation. https://en.wikipedia.org/wiki/Peppered_moth_evolution#Alternative_hypotheses. Accessed online on 9/14/2020
50. Stoye, Jonathan P. "Koala Retrovirus: a Genome Invasion in Real Time." *Genome Biology*, vol. 7, no. 11, 2006, doi:10.1186/gb-2006-7-11-241.
51. Tarlinton, Rachael E., et al. "Retroviral Invasion of the Koala Genome." *Nature*, vol. 442, no. 7098, 2006, pp. 79–81., doi:10.1038/nature04841.

52. Darwin, Charles. *On the Origin of Species by Means of Natural Selection, or Preservation of Favoured Races in the Struggle for Life.* John Murray, Albemarle Street, 1859, pp. 8-10. http://darwin-online.org.uk/converted/pdf/1859_Origin_F373.pdf
53. Phillips, David M, and Gil L Dryden. "Comparative Morphology of Mammalian Gametes." *A Comparative Overview of Mammalian Fertilization*, by Bonnie S. Dunbar and Michael G. O'Rand, Springer Science+Business Media, 2013.
54. Begun, David J, et al. "Evidence for De Novo Evolution of Testis-Expressed Genes in the Drosophila Yakuba/Drosophila Erecta Clade." *Genetics*, vol. 176, no. 2, 2007, pp. 1131–1137., doi:10.1534/genetics.106.069245.
55. Begun, David J, et al. "Recently Evolved Genes Identified from Drosophila Yakuba and D. Erecta Accessory Gland Expressed Sequence Tags." *Genetics*, vol. 172, no. 3, 2006, pp. 1675–1681., doi:10.1534/genetics.105.050336.
56. Levine, M. T., et al. "Novel Genes Derived from Noncoding DNA in Drosophila Melanogaster Are Frequently x-Linked and Exhibit Testis-Biased Expression." *Proceedings of the National Academy of Sciences*, vol. 103, no. 26, 2006, pp. 9935–9939., doi:10.1073/pnas.0509809103.
57. Voet, Donald, et al. *Fundamentals of Biochemistry: Life at the Molecular Level.* John Wiley & Sons, 2016, p. 929.
58. Lodish, Harvey. "Mobile DNA." *Molecular Cell Biology. 4th Edition.*, U.S. National Library of Medicine, 1 Jan. 1970. https://ncbi.nlm.nih.gov/books/NBK21495/. Accessed online on 1/25/2021
59. Wicker, Thomas, et al. "Impact of Transposable Elements on Genome Structure and Evolution in Bread Wheat." *Genome Biology*, vol. 19, no. 1, 2018, doi:10.1186/s13059-018-1479-0.
60. Pantzartzi, Chrysoula N., et al. "The Role of Transposable Elements in Functional Evolution of Amphioxus Genome: The Case of Opsin Gene Family." *Scientific Reports*, vol. 8, no. 1, 2018, doi:10.1038/s41598-018-20683-9.

61. Voet, Donald, et al. *Fundamentals of Biochemistry: Life at the Molecular Level.* John Wiley & Sons, 2016, p. 533
62. Britten, R. J., and E. H. Davidson. "Gene Regulation for Higher Cells: A Theory." *Science*, vol. 165, no. 3891, 1969, pp. 349–357., doi:10.1126/science.165.3891.349.
63. Britten, Roy J., and Eric H. Davidson. "Repetitive and Non-Repetitive DNA Sequences and a Speculation on the Origins of Evolutionary Novelty." *The Quarterly Review of Biology*, vol. 46, no. 2, 1971, pp. 111–138., doi:10.1086/406830.
64. King, M., and A. Wilson. "Evolution at Two Levels in Humans and Chimpanzees." *Science*, vol. 188, no. 4184, 1975, pp. 107–116., doi:10.1126/science.1090005.
65. Somel, M., et al. "Transcriptional Neoteny in the Human Brain." *Proceedings of the National Academy of Sciences*, vol. 106, no. 14, 2009, pp. 5743–5748., doi:10.1073/pnas.0900544106.
66. Studer, Anthony, et al. "Identification of a Functional Transposon Insertion in the Maize Domestication Gene TB1." *Nature Genetics*, vol. 43, no. 11, 2011, pp. 1160–1163., doi:10.1038/ng.942.
67. Bourque, G., et al. "Evolution of the Mammalian Transcription Factor Binding Repertoire via Transposable Elements." *Genome Research*, vol. 18, no. 11, 2008, pp. 1752–1762., doi:10.1101/gr.080663.108.
68. Lynch, Vincent J, et al. "Transposon-Mediated Rewiring of Gene Regulatory Networks Contributed to the Evolution of Pregnancy in Mammals." *Nature Genetics*, vol. 43, no. 11, 2011, pp. 1154–1159., doi:10.1038/ng.917.
69. Jordan, I.King, et al. "Origin of a Substantial Fraction of Human Regulatory Sequences from Transposable Elements." *Trends in Genetics*, vol. 19, no. 2, 2003, pp. 68–72., doi:10.1016/s0168-9525(02)00006-9.
70. Cain, Carolyn E, et al. "Gene Expression Differences among Primates Are Associated with Changes in a Histone Epigenetic Modification." *Genetics*, vol. 187, no. 4, 2011, pp. 1225–1234., doi:10.1534/genetics.110.126177.

71. Molaro, Antoine, et al. "Sperm Methylation Profiles Reveal Features of Epigenetic Inheritance and Evolution in Primates." *Cell*, vol. 146, no. 6, 2011, pp. 1029–1041., doi:10.1016/j.cell.2011.08.016.
72. Romero, Irene Gallego, et al. "Comparative Studies of Gene Expression and the Evolution of Gene Regulation." *Nature Reviews Genetics*, vol. 13, no. 7, 2012, pp. 505–516., doi:10.1038/nrg3229.
73. "Epigenomics Fact Sheet." *Genome.gov.* https://genome.gov/about-genomics/fact-sheets/Epigenomics-Fact-Sheet. Accessed online on 9/19/2021.
74. Bernstein, Bradley E., et al. "The Mammalian Epigenome." *Cell*, vol. 128, no. 4, 2007, pp. 669–681., doi:10.1016/j.cell.2007.01.033.
75. Castegna, Alessandra, et al. "The Mitochondrial Side of Epigenetics." *Physiological Genomics*, vol. 47, no. 8, 2015, pp. 299–307., doi:10.1152/physiolgenomics.00096.2014.
76. Martin, Antoine, et al. "A Transposon-Induced Epigenetic Change Leads to Sex Determination in Melon." *Nature*, vol. 461, no. 7267, 2009, pp. 1135–1138., doi:10.1038/nature08498.
77. Long, Manyuan, et al. "The Origin of New Genes: Glimpses from the Young and Old." *Nature Reviews Genetics*, vol. 4, no. 11, 2003, pp. 865–875., doi:10.1038/nrg1204.
78. Ohno, Susumu. *Evolution by Gene Duplication*. Springer, 1970.
79. Chandrasekaran, Chitra, and Esther Betrán. "Origins of New Genes and Pseudogenes." *Nature News*, Nature Publishing Group. https://nature.com/scitable/topicpage/origins-of-new-genes-and-pseudogenes-835/. Accessed online on 9/19/2021.
80. Magadum, Santoshkumar, et al. "Gene Duplication as a Major Force in Evolution." *Journal of Genetics*, vol. 92, no. 1, 2013, pp. 155–161., doi:10.1007/s12041-013-0212-8.
81. Nasvall, J., et al. "Real-Time Evolution of New Genes by Innovation, Amplification, and Divergence." *Science*, vol. 338, no. 6105, 2012, pp. 384–387., doi:10.1126/science.1226521.
82. Cerbin, Stefan, and Ning Jiang. "Duplication of Host Genes by Transposable Elements." *Current Opinion in Genetics & Development*, vol. 49, 2018, pp. 63–69., doi:10.1016/j.gde.2018.03.005.

83. Zhang, Jianbo, et al. "Generation of Tandem Direct Duplications by Reversed-Ends Transposition of Maize AC Elements." *PLoS Genetics*, vol. 9, no. 8, 2013, doi:10.1371/journal.pgen.1003691.
84. Kant, J. A., et al. "Evolution and Organization of the Fibrinogen Locus on Chromosome 4: Gene Duplication Accompanied by Transposition and Inversion." *Proceedings of the National Academy of Sciences*, vol. 82, no. 8, 1985, pp. 2344–2348., doi:10.1073/pnas.82.8.2344.
85. Morgante, Michele, et al. "Gene Duplication and Exon Shuffling by Helitron-like Transposons Generate Intraspecies Diversity in Maize." *Nature Genetics*, vol. 37, no. 9, 2005, pp. 997–1002., doi:10.1038/ng1615.
86. Chen, L., et al. "Evolution of Antifreeze Glycoprotein Gene from a Trypsinogen Gene in Antarctic Notothenioid Fish." *Proceedings of the National Academy of Sciences*, vol. 94, no. 8, 1997, pp. 3811–3816., doi:10.1073/pnas.94.8.3811.
87. Baalsrud, Helle Tessand, et al. "De Novo Gene Evolution of Antifreeze Glycoproteins in Codfishes Revealed by Whole Genome Sequence Data." *Molecular Biology and Evolution*, vol. 35, no. 3, 2017, pp. 593–606., doi:10.1093/molbev/msx311.
88. McLysaght, Aoife, and Daniele Guerzoni. "New Genes from Non-coding Sequence: The Role of De Novo Protein-Coding Genes in Eukaryotic Evolutionary Innovation." *Philosophical Transactions of the Royal Society B: Biological Sciences*, vol. 370, no. 1678, 2015, p. 20140332., doi:10.1098/rstb.2014.0332.
89. Palmieri, Nicola, et al. "The Life Cycle of Drosophila Orphan Genes." *ELife*, vol. 3, 2014, doi:10.7554/elife.01311.
90. Toll-Riera, M., et al. "Origin of Primate Orphan Genes: A Comparative Genomics Approach." *Molecular Biology and Evolution*, vol. 26, no. 3, 2008, pp. 603–612., doi:10.1093/molbev/msn281.
91. Chen, Shou-Tao, et al. "Evolution of Hydra, a Recently Evolved Testis-Expressed Gene with Nine Alternative First Exons in Drosophila Melanogaster." *PLoS Genetics*, preprint, no. 2007, 2005, doi:10.1371/journal.pgen.0030107.eor.
92. Jin, Gui-Hua, et al. "Genetic Innovations: Transposable Element Recruitment and De Novo Formation Lead to the Birth of Orphan

Genes in the Rice Genome." *Journal of Systematics and Evolution*, vol. 59, no. 2, 2019, pp. 341–351., doi:10.1111/jse.12548.

93. Liu, Yang, et al. "Structure and Evolutionary Origin of ca2+-Dependent Herring Type II Antifreeze Protein." *PLoS ONE*, vol. 2, no. 6, 2007, doi:10.1371/journal.pone.0000548.
94. Graham, Laurie A., et al. "Helical Antifreeze Proteins Have Independently Evolved in Fishes on Four Occasions." *PLoS ONE*, vol. 8, no. 12, 2013, doi:10.1371/journal.pone.0081285.
95. Britten, R. J. "Coding Sequences of Functioning Human Genes Derived Entirely from Mobile Element Sequences." *Proceedings of the National Academy of Sciences*, vol. 101, no. 48, 2004, pp. 16825–16830., doi:10.1073/pnas.0406985101.
96. Blond, Jean-Luc, et al. "Molecular Characterization and Placental Expression of HERV-w, a New Human Endogenous Retrovirus Family." *Journal of Virology*, vol. 73, no. 2, 1999, pp. 1175–1185., doi:10.1128/jvi.73.2.1175-1185.1999.
97. Mi, Sha, et al. "Syncytin Is a Captive Retroviral Envelope Protein Involved in Human Placental Morphogenesis." *Nature*, vol. 403, no. 6771, 2000, pp. 785–789., doi:10.1038/35001608.
98. Vargas, Amandine, et al. "Syncytin-2 Plays an Important Role in the Fusion of Human Trophoblast Cells." *Journal of Molecular Biology*, vol. 392, no. 2, 2009, pp. 301–318., doi:10.1016/j.jmb.2009.07.025.
99. Nekrutenko, Anton, and Wen-Hsiung Li. "Transposable Elements Are Found in a Large Number of Human Protein-Coding Genes." *Trends in Genetics*, vol. 17, no. 11, 2001, pp. 619–621., doi:10.1016/s0168-9525(01)02445-3.
100. Mojica, F. J., et al. "Transcription at Different Salinities of Haloferax Mediterranei Sequences Adjacent to Partially Modified Psti Sites." *Molecular Microbiology*, vol. 9, no. 3, 1993, pp. 613–621., doi:10.1111/j.1365-2958.1993.tb01721.x.
101. Jansen, Ruud., et al. "Identification of Genes That Are Associated with DNA Repeats in Prokaryotes." *Molecular Microbiology*, vol. 43, no. 6, 2002, pp. 1565–1575., doi:10.1046/j.1365-2958.2002.02839.x.

102. Lander, Eric S. "The Heroes of CRISPR." *Cell*, vol. 164, no. 1-2, 2016, pp. 18–28., doi:10.1016/j.cell.2015.12.041.
103. Jinek, Martin, et al. "A Programmable Dual-RNA–Guided DNA Endonuclease in Adaptive Bacterial Immunity." *Science*, vol. 337, no. 6096, 2012, pp. 816–821., doi:10.1126/science.1225829.
104. Gasiunas, Giedrius, et al. "Cas9–Crrna Ribonucleoprotein Complex Mediates Specific DNA Cleavage for Adaptive Immunity in Bacteria." *Proceedings of the National Academy of Sciences*, vol. 109, no. 39, 2012, doi:10.1073/pnas.1208507109.
105. Mali, P., et al. "RNA-Guided Human Genome Engineering via cas9." *Science*, vol. 339, no. 6121, 2013, pp. 823–826., doi:10.1126/science.1232033.
106. Cong, L., et al. "Multiplex Genome Engineering Using CRISPR/Cas Systems." *Science*, vol. 339, no. 6121, 2013, pp. 819–823., doi:10.1126/science.1231143.
107. Jinek, Martin, et al. "RNA-Programmed Genome Editing in Human Cells." *ELife*, vol. 2, 2013, doi:10.7554/elife.00471.
108. Cho, Seung Woo, et al. "Targeted Genome Engineering in Human Cells with the cas9 RNA-Guided Endonuclease." *Nature Biotechnology*, vol. 31, no. 3, 2013, pp. 230–232., doi:10.1038/nbt.2507.
109. Hwang, Woong Y, et al. "Efficient Genome Editing in Zebrafish Using a CRISPR-Cas System." *Nature Biotechnology*, vol. 31, no. 3, 2013, pp. 227–229., doi:10.1038/nbt.2501.

Chapter 6

1. Woese, C. R., and G. E. Fox. "Phylogenetic Structure of the Prokaryotic Domain: The Primary Kingdoms." *Proceedings of the National Academy of Sciences*, vol. 74, no. 11, 1977, pp. 5088–5090., doi:10.1073/pnas.74.11.5088.
2. Woese, C. R., et al. "Towards a Natural System of Organisms: Proposal for the Domains Archaea, Bacteria, and Eucarya." *Proceedings of the National Academy of Sciences*, vol. 87, no. 12, 1990, pp. 4576–4579., doi:10.1073/pnas.87.12.4576.

3. Acheson, Nicholas H. *Fundamentals of Molecular Virology.* Wiley, 2011.
4. Desjardins, Christopher A. "Unusual Viral Genomes." *Mimivirus - an Overview | ScienceDirect Topics.* https://sciencedirect.com/topics/medicine-and-dentistry/mimivirus. Accessed online on 9/30/2021.
5. Rybicki , Edward P, and Russell Kightley. "A Short History of the Discovery of Viruses." *ViroBlogy.* https://rybicki.blog/tag/chamberland-filter/. Accessed online on 9/30/2021.
6. OpenStax. "Viruses." *Allied Health Microbiology*, Oregon State University. https://open.oregonstate.education/microbiology/chapter/6-1viruses/. Accessed online on 9/30/2021.
7. Foster, Jerome E, and Gustavo Fermin. "Origin and Evolution of Viruses." *Viruses Molecular Biology, Host Interactions and Applications to Biotechnology*, by Paula Tennant et al., Academic Press, 2018, p. 85.
8. Forterre, Patrick. "The Origin of Viruses and Their Possible Roles in Major Evolutionary Transitions." *Virus Research*, vol. 117, no. 1, 2006, pp. 5–16., doi:10.1016/j.virusres.2006.01.010.
9. Forterre, Patrick, et al. "Origin and Evolution of DNA and DNA Replication Machineries." *Madame Curie Bioscience Database [Internet].*, U.S. National Library of Medicine, 1 Jan. 1970. https://ncbi.nlm.nih.gov/books/NBK6360/. Accessed online 3/12/2021
10. Jardine, Paul J, and Dwight L Anderson. "DNA Packaging in Double-Stranded DNA Phages." *The Bacteriophages*, by Richard Calendar, Oxford University Press, 2006, p. 55.
11. Blanco, Antonio, and Gustavo Blanco. "Nucleic Acids." *Medical Biochemistry*, 2017, pp. 121–140., doi:10.1016/b978-0-12-803550-4.00006-9.
12. Voet, Donald, et al. *Fundamentals of Biochemistry: Life at the Molecular Level.* John Wiley & Sons, 2016, pp: 44-46
13. Scofield, Margaret. "Nucleic Acids." *XPharm: The Comprehensive Pharmacology Reference*, 2007, pp. 1–15., doi:10.1016/b978-008055232-3.60059-5.
14. Joyce, Gerald F. "RNA Evolution and the Origins of Life." *Nature*, vol. 338, no. 6212, 1989, pp. 217–224., doi:10.1038/338217a0.

15. Hazen, R. M., and D. A. Sverjensky. "Mineral Surfaces, Geochemical Complexities, and the Origins of Life." *Cold Spring Harbor Perspectives in Biology*, vol. 2, no. 5, 2010, doi:10.1101/cshperspect.a002162.
16. Maurel, Marie-Christine, and Fabrice Leclerc. "From Foundation Stones to Life: Concepts and Results." *Elements*, vol. 12, no. 6, 2016, pp. 407–412., doi:10.2113/gselements.12.6.407.
17. Biondi, Elisa, et al. "Catalytic Activity of Hammerhead Ribozymes in a Clay Mineral Environment: Implications for the RNA World." *Gene*, vol. 389, no. 1, 2007, pp. 10–18., doi:10.1016/j.gene.2006.09.002.
18. Ferris, James P., et al. "Synthesis of Long Prebiotic Oligomers on Mineral Surfaces." *Nature*, vol. 381, no. 6577, 1996, pp. 59–61., doi:10.1038/381059a0.
19. Ferris, J. P. "Mineral Catalysis and Prebiotic Synthesis: Montmorillonite-Catalyzed Formation of RNA." *Elements*, vol. 1, no. 3, 2005, pp. 145–149., doi:10.2113/gselements.1.3.145.
20. Šponer, Judit E., et al. "Emergence of the First Catalytic Oligonucleotides in a Formamide-Based Origin Scenario." *Chemistry - A European Journal*, vol. 22, no. 11, 2016, pp. 3572–3586., doi:10.1002/chem.201503906.
21. Costanzo, Giovanna, et al. "Nucleoside Phosphorylation by Phosphate Minerals." *Journal of Biological Chemistry*, vol. 282, no. 23, 2007, pp. 16729–16735., doi:10.1074/jbc.m611346200.

Chapter 7

1. Voet, Donald, et al. *Fundamentals of Biochemistry: Life at the Molecular Level.* John Wiley & Sons, 2016, p. 2.
2. Hazen, Robert M. *The Story of Earth: The First 4.5 Billion Years, from Stardust to Living Planet.* Viking, 2013, p. 5.
3. Hazen, Robert M. *The Story of Earth: The First 4.5 Billion Years, from Stardust to Living Planet.* Viking, 2013, pp. 11-12.
4. Hazen, Robert M. *The Story of Earth: The First 4.5 Billion Years, from Stardust to Living Planet.* Viking, 2013, p. 135.
5. "Darwin Correspondence Project." *Darwin Correspondence Project.* https://darwinproject.ac.uk/letter/DCP-LETT-7471.xml. Accessed online on 9/27/2020.

6. Da Silva, Laura, et al. "Salt-Promoted Synthesis of RNA-like Molecules in Simulated Hydrothermal Conditions." *Journal of Molecular Evolution*, vol. 80, no. 2, 2014, pp. 86–97., doi:10.1007/s00239-014-9661-9.
7. Martins, Zita. "The Nitrogen Heterocycle Content of Meteorites and Their Significance for the Origin of Life." *Life*, vol. 8, no. 3, 2018, p. 28., doi:10.3390/life8030028.
8. Glavin, D P, and J L Bada. *Isolation of Purines and Pyrimidines from the Murchison Meteorite Using Sublimation*. https://www.lpi.usra.edu/meetings/lpsc2004/pdf/1022.pdf. Accessed online on 03/20/2022.
9. Rotelli, Luca, et al. "The Key Role of Meteorites in the Formation of Relevant Prebiotic Molecules in a Formamide/Water Environment." *Scientific Reports*, vol. 6, no. 1, 2016, doi:10.1038/srep38888.
10. Saladino, R. "A Possible Prebiotic Synthesis of Purine, Adenine, Cytosine, and 4(3H)-Pyrimidinone from Formamide Implications for the Origin of Life." *Bioorganic & Medicinal Chemistry*, vol. 9, no. 5, 2001, pp. 1249–1253., doi:10.1016/s0968-0896(00)00340-0.
11. Saladino, Raffaele, et al. "Meteorite-Catalyzed Syntheses of Nucleosides and of Other Prebiotic Compounds from Formamide under Proton Irradiation." *Proceedings of the National Academy of Sciences*, vol. 112, no. 21, 2015, doi:10.1073/pnas.1422225112.
12. Saladino, Raffaele, et al. "Meteorites as Catalysts for Prebiotic Chemistry." *Chemistry - A European Journal*, vol. 19, no. 50, 2013, pp. 16916–16922., doi:10.1002/chem.201303690.
13. Ferus, Martin, et al. "High-Energy Chemistry of Formamide: A Unified Mechanism of Nucleobase Formation." *Proceedings of the National Academy of Sciences*, vol. 112, no. 3, 2014, pp. 657–662., doi:10.1073/pnas.1412072111.
14. Šponer, Judit E., et al. "Emergence of the First Catalytic Oligonucleotides in a Formamide-Based Origin Scenario." *Chemistry - A European Journal*, vol. 22, no. 11, 2016, pp. 3572–3586., doi:10.1002/chem.201503906.
15. Saladino, Raffaele, et al. "Formamide and the Origin of Life." *Physics of Life Reviews*, vol. 9, no. 1, 2012, pp. 84–104., doi:10.1016/j.plrev.2011.12.002.

16. Barks, Hannah L., et al. "Guanine, Adenine, and Hypoxanthine Production in UV-Irradiated Formamide Solutions: Relaxation of the Requirements for Prebiotic Purine Nucleobase Formation." *ChemBioChem*, vol. 11, no. 9, 2010, pp. 1240–1243., doi:10.1002/cbic.201000074.
17. Costanzo, Giovanna, et al. "Nucleoside Phosphorylation by Phosphate Minerals." *Journal of Biological Chemistry*, vol. 282, no. 23, 2007, pp. 16729–16735., doi:10.1074/jbc.m611346200.
18. Dibrova, Daria V., et al. "The Role of Energy in the Emergence of Biology from Chemistry." *Origins of Life and Evolution of Biospheres*, vol. 42, no. 5, 2012, pp. 459–468., doi:10.1007/s11084-012-9308-z.
19. Ferris, J. P. "Mineral Catalysis and Prebiotic Synthesis: Montmorillonite-Catalyzed Formation of RNA." *Elements*, vol. 1, no. 3, 2005, pp. 145–149., doi:10.2113/gselements.1.3.145.

Chapter 8

1. Rotelli, Luca, et al. "The Key Role of Meteorites in the Formation of Relevant Prebiotic Molecules in a Formamide/Water Environment." *Scientific Reports*, vol. 6, no. 1, 2016, doi:10.1038/srep38888.
2. Saladino, Raffaele, et al. "Meteorites as Catalysts for Prebiotic Chemistry." *Chemistry - A European Journal*, vol. 19, no. 50, 2013, pp. 16916–16922., doi:10.1002/chem.201303690.
3. Saladino, Raffaele, et al. "Meteorite-Catalyzed Syntheses of Nucleosides and of Other Prebiotic Compounds from Formamide under Proton Irradiation." *Proceedings of the National Academy of Sciences*, vol. 112, no. 21, 2015, doi:10.1073/pnas.1422225112.
4. Belmonte, Luca, and Sheref S. Mansy. "Metal Catalysts and the Origin of Life." *Elements*, vol. 12, no. 6, 2016, pp. 413–418., doi:10.2113/gselements.12.6.413.
5. Tock, Mark R., et al. "Dynamic Evidence for Metal Ion Catalysis in the Reaction Mediated by a Flap Endonuclease." *The EMBO Journal*, vol. 22, no. 5, 2003, pp. 995–1004., doi:10.1093/emboj/cdg098.
6. Yang, Wei, et al. "Making and Breaking Nucleic Acids: Two-Mg2+-Ion Catalysis and Substrate Specificity." *Molecular Cell*, vol. 22, no. 1, 2006, pp. 5–13., doi:10.1016/j.molcel.2006.03.013.

7. Symons, R. H., and J. W. Randles. "Encapsidated Circular Viroid-like Satellite RNAS (Virusoids) of Plants." *Current Topics in Microbiology and Immunology*, 1999, pp. 81–105., doi:10.1007/978-3-662-09796-0_5.
8. Kanavarioti, Anastassia, et al. "Magnesium Ion Catalyzed p-N Bond Hydrolysis in Imidazolide Activated Nucleotides. Relevance to Template-Directed Synthesis of Polynucleotides." *Origins of Life and Evolution of the Biosphere*, vol. 19, no. 3-5, 1989, pp. 353–353., doi:10.1007/bf02388889.
9. Sissi, C., and M. Palumbo. "Effects of Magnesium and Related Divalent Metal Ions in Topoisomerase Structure and Function." *Nucleic Acids Research*, vol. 37, no. 3, 2009, pp. 702–711., doi:10.1093/nar/gkp024.
10. Voet, Donald, et al. *Fundamentals of Biochemistry: Life at the Molecular Level.* John Wiley & Sons, 2016, pp. 975-976.
11. Theng, Benny K.G. "Preface." *Clay Mineral Catalysis of Organic Reactions*, CRC PRESS, 2020.
12. Nelson, Stephen A. *Weathering & Clay Minerals.* https://tulane.edu/~sanelson/eens211/weathering&clayminerals. Accessed online on 4/9/2021
13. Hashizume, Hideo. "Role of Clay Minerals in Chemical Evolution and the Origins of Life." *Clay Minerals in Nature - Their Characterization, Modification and Application*, 2012, doi:10.5772/50172.
14. Ferris, J. P. "Mineral Catalysis and Prebiotic Synthesis: Montmorillonite-Catalyzed Formation of RNA." *Elements*, vol. 1, no. 3, 2005, pp. 145–149., doi:10.2113/gselements.1.3.145.
15. Saladino, Raffaele, et al. "Synthesis and Degradation of Nucleobases and Nucleic Acids by Formamide in the Presence of Montmorillonites." *ChemBioChem*, vol. 5, no. 11, 2004, pp. 1558–1566., doi:10.1002/cbic.200400119.
16. Costanzo, Giovanna, et al. "Formamide as the Main Building Block in the Origin of Nucleic Acids." *BMC Evolutionary Biology*, vol. 7, no. S2, 2007, doi:10.1186/1471-2148-7-s2-s1.
17. Hashizume, Hideo. "Adsorption of Nucleic Acid Bases, Ribose, and Phosphate by Some Clay Minerals." *Life*, vol. 5, no. 1, 2015, pp. 637–650., doi:10.3390/life5010637.

18. Ferris, James P., et al. "Synthesis of Long Prebiotic Oligomers on Mineral Surfaces." *Nature*, vol. 381, no. 6577, 1996, pp. 59–61., doi:10.1038/381059a0.
19. Ferris, James P. "Montmorillonite-Catalysed Formation of RNA Oligomers: The Possible Role of Catalysis in the Origins of Life." *Philosophical Transactions of the Royal Society B: Biological Sciences*, vol. 361, no. 1474, 2006, pp. 1777–1786., doi:10.1098/rstb.2006.1903.
20. "Humpty Dumpty." *All Nursery Rhymes*. https://allnurseryrhymes.com/humpty-dumpty/. Accessed online on 10/2/2020.
21. Miller, S. L. "A Production of Amino Acids under Possible Primitive Earth Conditions." *Science*, vol. 117, no. 3046, 1953, pp. 528–529., doi:10.1126/science.117.3046.528.
22. Kaufmann, Michael. "On the Free Energy That Drove Primordial Anabolism." *International Journal of Molecular Sciences*, vol. 10, no. 4, 2009, pp. 1853–1871., doi:10.3390/ijms10041853.
23. Chyba, Christopher, and Carl Sagan. "Endogenous Production, Exogenous Delivery and Impact-Shock Synthesis of Organic Molecules: An Inventory for the Origins of Life." *Nature*, vol. 355, no. 6356, 1992, pp. 125–132., doi:10.1038/355125a0.
24. Ferus, Martin, et al. "High-Energy Chemistry of Formamide: A Unified Mechanism of Nucleobase Formation." *Proceedings of the National Academy of Sciences*, vol. 112, no. 3, 2014, pp. 657–662., doi:10.1073/pnas.1412072111.
25. Muller, Anthonie W., and Dirk Schulze-Makuch. "Thermal Energy and the Origin of Life." *Origins of Life and Evolution of Biospheres*, vol. 36, no. 2, 2006, pp. 177–189., doi:10.1007/s11084-005-9003-4.
26. Costanzo, Giovanna, et al. "Nucleoside Phosphorylation by Phosphate Minerals." *Journal of Biological Chemistry*, vol. 282, no. 23, 2007, pp. 16729–16735., doi:10.1074/jbc.m611346200.
27. Da Silva, Laura, et al. "Salt-Promoted Synthesis of RNA-like Molecules in Simulated Hydrothermal Conditions." *Journal of Molecular Evolution*, vol. 80, no. 2, 2014, pp. 86–97., doi:10.1007/s00239-014-9661-9.

28. Stetter, Karl O. "Hyperthermophiles in the History of Life." *Philosophical Transactions of the Royal Society B: Biological Sciences*, vol. 361, no. 1474, 2006, pp. 1837–1843., doi:10.1098/rstb.2006.1907.
29. Van, Dover Cindy Lee. *The Ecology of Deep-Sea Hydrothermal Vents*. Princeton University Press, 2000, p. 117.
30. Chi, Andrew, and Robert G. Kemp. "The Primordial High Energy Compound: ATP or Inorganic Pyrophosphate?" *Journal of Biological Chemistry*, vol. 275, no. 46, 2000, pp. 35677–35679., doi:10.1074/jbc.c000581200.
31. PONNAMPERUMA, CYRIL, et al. "Synthesis of Adenosine Triphosphate under Possible Primitive Earth Conditions." *Nature*, vol. 199, no. 4890, 1963, pp. 222–226., doi:10.1038/199222a0.
32. Barks, Hannah L., et al. "Guanine, Adenine, and Hypoxanthine Production in UV-Irradiated Formamide Solutions: Relaxation of the Requirements for Prebiotic Purine Nucleobase Formation." *ChemBioChem*, vol. 11, no. 9, 2010, pp. 1240–1243., doi:10.1002/cbic.201000074.
33. Simoncini, E., et al. "Modeling Free Energy Availability from Hadean Hydrothermal Systems to the First Metabolism." *Origins of Life and Evolution of Biospheres*, vol. 41, no. 6, 2011, pp. 529–532., doi:10.1007/s11084-011-9251-4.
34. Simoncini , E, et al. "Thermodynamics of Chemical Free Energy Generation in off-Axis Hydrothermal Vent Systems and Its Consequences for Compartmentalization and the Emergence of Life." *Arxiv*, 12 Aug. 2010. https://arxiv.org/ftp/arxiv/papers/1008/1008.2176.pdf

Chapter 9

1. Ferris, James P., et al. "Synthesis of Long Prebiotic Oligomers on Mineral Surfaces." *Nature*, vol. 381, no. 6577, 1996, pp. 59–61., doi:10.1038/381059a0.
2. Ferris, J. P. "Mineral Catalysis and Prebiotic Synthesis: Montmorillonite-Catalyzed Formation of RNA." *Elements*, vol. 1, no. 3, 2005, pp. 145–149., doi:10.2113/gselements.1.3.145.

3. Ertem, Gözen, and James P. Ferris. "Sequence- and Regio-Selectivity in the Montmorillonite-Catalyzed Synthesis of RNA." *Origins of Life and Evolution of the Biosphere*, vol. 30, no. 5, 2000, pp. 411–422., doi:10.1023/a:1006767019897.
4. Miyakawa, Shin, and James P. Ferris. "Sequence- and Regioselectivity in the Montmorillonite-Catalyzed Synthesis of RNA." *Journal of the American Chemical Society*, vol. 125, no. 27, 2003, pp. 8202–8208., doi:10.1021/ja034328e.
5. Dibrova, Daria V., et al. "The Role of Energy in the Emergence of Biology from Chemistry." *Origins of Life and Evolution of Biospheres*, vol. 42, no. 5, 2012, pp. 459–468., doi:10.1007/s11084-012-9308-z.
6. Costanzo, Giovanna, et al. "Nucleoside Phosphorylation by Phosphate Minerals." *Journal of Biological Chemistry*, vol. 282, no. 23, 2007, pp. 16729–16735., doi:10.1074/jbc.m611346200.
7. Saladino, Raffaele, et al. "Formamide and the Origin of Life." *Physics of Life Reviews*, vol. 9, no. 1, 2012, pp. 84–104., doi:10.1016/j.plrev.2011.12.002.
8. Šponer, Judit E., et al. "Emergence of the First Catalytic Oligonucleotides in a Formamide-Based Origin Scenario." *Chemistry - A European Journal*, vol. 22, no. 11, 2016, pp. 3572–3586., doi:10.1002/chem.201503906.
9. Roossinck, M J, et al. "Satellite RNAS of Plant Viruses: Structures and Biological Effects." *Microbiological Reviews*, vol. 56, no. 2, 1992, pp. 265–279., doi:10.1128/mr.56.2.265-279.1992.
10. Symons, R. H., and J. W. Randles. "Encapsidated Circular Viroid-like Satellite RNAS (Virusoids) of Plants." *Current Topics in Microbiology and Immunology*, 1999, pp. 81–105., doi:10.1007/978-3-662-09796-0_5.
11. Lodish, Harvey. *Molecular Cell Biology*. W.H. Freeman and Company, 2016, pp. 193-195
12. Voet, Donald, et al. *Fundamentals of Biochemistry: Life at the Molecular Level*. John Wiley & Sons, 2016, pp: 1006-1024.
13. Joyce, Gerald F. "RNA Evolution and the Origins of Life." *Nature*, vol. 338, no. 6212, 1989, pp. 217–224., doi:10.1038/338217a0.

14. Vaghefi, Morteza. "Chemical Synthesis of Nucleotide Triphosphates." *Nucleoside Triphosphates and Their Analogs Chemistry, Biotechnology, and Biological Applications*, by Seyed-Morteza Monir-Vaghefi, Taylor & Francis, 2005, p. 106.
15. Chi, Andrew, and Robert G. Kemp. "The Primordial High Energy Compound: ATP or Inorganic Pyrophosphate?" *Journal of Biological Chemistry*, vol. 275, no. 46, 2000, pp. 35677–35679., doi:10.1074/jbc.c000581200.
16. PONNAMPERUMA, CYRIL, et al. "Synthesis of Adenosine Triphosphate under Possible Primitive Earth Conditions." *Nature*, vol. 199, no. 4890, 1963, pp. 222–226., doi:10.1038/199222a0.
17. Barbatti, Mario, et al. "The UV Absorption of Nucleobases: Semi-Classical Ab Initio Spectra Simulations." *Physical Chemistry Chemical Physics*, vol. 12, no. 19, 2010, p. 4959., doi:10.1039/b924956g.
18. Vorlíčková, M., et al. "Nucleic Acids: Spectroscopic Methods." *Encyclopedia of Analytical Science (Second Edition)*, Elsevier, 28 May 2005, https://sciencedirect.com/science/article/pii/B0123693977004283
19. Voet, Donald, et al. *Fundamentals of Biochemistry: Life at the Molecular Level.* John Wiley & Sons, 2016, p. 884.
20. Blackburn, Elizabeth H. "Structure and Function of Telomeres." *Nature*, vol. 350, no. 6319, 1991, pp. 569–573., doi:10.1038/350569a0.
21. Navarro, J.-A. "Characterization of the Initiation Sites of Both Polarity Strands of a Viroid RNA Reveals a Motif Conserved in Sequence and Structure." *The EMBO Journal*, vol. 19, no. 11, 2000, pp. 2662–2670., doi:10.1093/emboj/19.11.2662.
22. Delgado, Sonia, et al. "A Short Double-Stranded RNA Motif of Peach Latent Mosaic Viroid Contains the Initiation and the Self-Cleavage Sites of Both Polarity Strands." *Journal of Virology*, vol. 79, no. 20, 2005, pp. 12934–12943., doi:10.1128/jvi.79.20.12934-12943.2005.
23. Auchincloss, Andrea H., and Gregory G. Brown. "Soybean Mitochondrial Transcripts Capped in Vitro with Guanylyltransferase." *Biochemistry and Cell Biology*, vol. 67, no. 6, 1989, pp. 315–319., doi:10.1139/o89-049.

24. REPSILBER, DIRK, et al. "Formation of Metastable RNA Structures by Sequential Folding during Transcription: Time-Resolved Structural Analysis of Potato Spindle Tuber Viroid (-)-Stranded RNA by Temperature-Gradient Gel Electrophoresis." *RNA*, vol. 5, no. 4, 1999, pp. 574–584., doi:10.1017/s1355838299982018.
25. Lodish, Harvey. *Molecular Cell Biology*. W.H. Freeman and Company, 2016, pp: 244-245.
26. Paul, N., and G. F. Joyce. "A Self-Replicating Ligase Ribozyme." *Proceedings of the National Academy of Sciences*, vol. 99, no. 20, 2002, pp. 12733–12740., doi:10.1073/pnas.202471099.
27. Wochner, A., et al. "Ribozyme-Catalyzed Transcription of an Active Ribozyme." *Science*, vol. 332, no. 6026, 2011, pp. 209–212., doi:10.1126/science.1200752.
28. Johnston, Wendy K., et al. "RNA-Catalyzed RNA Polymerization: Accurate and General RNA-Templated Primer Extension." *Science*, vol. 292, no. 5520, 2001, pp. 1319–1325., doi:10.1126/science.1060786.
29. Bregestovski, P. D. "'Rna World', a Highly Improbable Scenario of the Origin and Early Evolution of Life on Earth." *Journal of Evolutionary Biochemistry and Physiology*, vol. 51, no. 1, 2015, pp. 72–84., doi:10.1134/s0022093015010111.
30. Brovarone, Alberto Vitale, et al. "Let There Be Water: How Hydration/Dehydration Reactions Accompany Key Earth and Life Processes#." *American Mineralogist*, vol. 105, no. 8, 2020, pp. 1152–1160., doi:10.2138/am-2020-7380.
31. Damer, Bruce, and David Deamer. "The Hot Spring Hypothesis for an Origin of Life." *Astrobiology*, vol. 20, no. 4, 2020, pp. 429–452., doi:10.1089/ast.2019.2045.
32. Biondi, Elisa, et al. "Catalytic Activity of Hammerhead Ribozymes in a Clay Mineral Environment: Implications for the RNA World." *Gene*, vol. 389, no. 1, 2007, pp. 10–18., doi:10.1016/j.gene.2006.09.002.
33. Ferris, James P. "Montmorillonite-Catalysed Formation of RNA Oligomers: The Possible Role of Catalysis in the Origins of Life." *Philosophical Transactions of the Royal Society B: Biological Sciences*, vol. 361, no. 1474, 2006, pp. 1777–1786., doi:10.1098/rstb.2006.1903.

34. Kitch, William. "CE 531 Mod 2.1.1: Clay Mineralogy." *YouTube*, YouTube, 9 Oct. 2014, https://youtube.com/watch?v=VqCQDpjdKKE. Viewed online on 4/12/2021.
35. Hazen, R. M., and D. A. Sverjensky. "Mineral Surfaces, Geochemical Complexities, and the Origins of Life." *Cold Spring Harbor Perspectives in Biology*, vol. 2, no. 5, 2010, doi:10.1101/cshperspect.a002162.
36. Miller, Stanley L. "A Production of Amino Acids under Possible Primitive Earth Conditions." *Science*, vol. 117, no. 3046, 1953, pp. 528–529., doi:10.1126/science.117.3046.528.
37. Rotelli, Luca, et al. "The Key Role of Meteorites in the Formation of Relevant Prebiotic Molecules in a Formamide/Water Environment." *Scientific Reports*, vol. 6, no. 1, 2016, doi:10.1038/srep38888.
38. Theng, Benny K.G. *Clay Mineral Catalysis of Organic Reactions*. CRC PRESS, 2019, p.398.
39. Saladino, Raffaele, et al. "Synthesis and Degradation of Nucleobases and Nucleic Acids by Formamide in the Presence of Montmorillonites." *ChemBioChem*, vol. 5, no. 11, 2004, pp. 1558–1566., doi:10.1002/cbic.200400119.
40. Costanzo, Giovanna, et al. "Formamide as the Main Building Block in the Origin of Nucleic Acids." *BMC Evolutionary Biology*, vol. 7, no. S2, 2007, doi:10.1186/1471-2148-7-s2-s1.
41. Hashizume, Hideo. "Role of Clay Minerals in Chemical Evolution and the Origins of Life." *Clay Minerals in Nature - Their Characterization, Modification and Application*, 2012, doi:10.5772/50172.
42. Erastova, Valentina, et al. "Mineral Surface Chemistry Control for Origin of Prebiotic Peptides." *Nature Communications*, vol. 8, no. 1, 2017, doi:10.1038/s41467-017-02248-y.
43. Deamer, David W. "Hydrothermal Conditions Are Conducive for the Origin of Life." *Assembling Life*, 2019, doi:10.1093/oso/9780190646387.003.0008.
44. Bolik, S., et al. "First Experimental Evidence for the Preferential Stabilization of the Natural D- over the Nonnatural L-Configuration in Nucleic Acids." *RNA*, vol. 13, no. 11, 2007, pp. 1877–1880., doi:10.1261/rna.564507.

45. Fujii, Noriko. "D-Amino Acids in Living Higher Organisms." *Origins of Life and Evolution of the Biosphere*, vol. 32, no. 2, 2002, pp. 103–127., doi:10.1023/a:1016031014871.
46. Fujii, Noriko, et al. "D-Amino Acids in Protein: The Mirror of Life as a Molecular Index of Aging." *Biochimica Et Biophysica Acta (BBA) - Proteins and Proteomics*, vol. 1866, no. 7, 2018, pp. 840–847., doi:10.1016/j.bbapap.2018.03.001.
47. Sachs-Barrable, Kristina, et al. "The Effect of Two Novel Cholesterol-Lowering Agents, Disodium Ascorbyl Phytostanol Phosphate (DAPP) and Nanostructured Aluminosilicate (Nsas) on the Expression and Activity of P-Glycoprotein within Caco-2 Cells." *Lipids in Health and Disease,* vol. 13, no. 1, 2014, doi:10.1186/1476-511x-13-153.
48. Lee, Nick, et al. "Ribozyme-Catalyzed TRNA Aminoacylation." *Nature Structural Biology*, vol. 7, no. 1, 2000, pp. 28–33., doi:10.1038/71225.
49. Sabate, Raimon, et al. "Protein Folding and Aggregation in Bacteria." *Cellular and Molecular Life Sciences*, vol. 67, no. 16, 2010, pp. 2695–2715., doi:10.1007/s00018-010-0344-4.
50. Seddighi, Hassan, et al. "LDH(Mg/Al:2)@Montmorillonite Nanocomposite as a Novel Anion-Exchanger to Adsorb Uranyl Ion from Carbonate-Containing Solutions." *Journal of Radioanalytical and Nuclear Chemistry*, vol. 314, no. 1, 2017, pp. 415–427., doi:10.1007/s10967-017-5387-7.
51. Yi, Deqi, et al. "Layered Double Hydroxide - Montmorillonite - a New Nano-Dimensional Material." *Polymers for Advanced Technologies*, vol. 24, no. 2, 2012, pp. 204–209., doi:10.1002/pat.3072.
52. Makwana, Dipak, et al. "MG-FE Layered Double Hydroxide Modified Montmorillonite as Hydrophilic Nanofiller in Polysulfone-Polyvinylpyrrolidone Blend Ultrafiltration Membranes: Separation of Oil-Water Mixture." *Applied Clay Science*, vol. 192, 2020, p. 105636., doi:10.1016/j.clay.2020.105636.
53. "Interstellar Molecules - the Cosmic Ice Laboratory - Sciences and Exploration Directorate - NASA's Goddard Space Flight Center."

NASA, NASA. https://science.gsfc.nasa.gov/691/cosmicice/interstellar.html. Accessed online on 3/30/2021

54. Shatilovich, A. V., et al. "Viable Nematodes from Late Pleistocene Permafrost of the Kolyma River Lowland." *Doklady Biological Sciences*, vol. 480, no. 1, 2018, pp. 100–102., doi:10.1134/s0012496618030079.
55. Vreeland, Russell H., et al. "Isolation of a 250 Million-Year-Old Halotolerant Bacterium from a Primary Salt Crystal." *Nature*, vol. 407, no. 6806, 2000, pp. 897–900., doi:10.1038/35038060.
56. Cano, R., and M. Borucki. "Revival and Identification of Bacterial Spores in 25- to 40-Million-Year-Old Dominican Amber." *Science*, vol. 268, no. 5213, 1995, pp. 1060–1064., doi:10.1126/science.7538699.
57. Kumar, Sangeetha Devi, et al. "Evidence for Low-Level Translation in Human Erythrocytes." *Molecular Biology of the Cell,* vol. 33, no. 12, 2022, doi:10.1091/mbc.e21-09-0437.
58. Nissen, Poul, et al. "The Structural Basis of Ribosome Activity in Peptide Bond Synthesis." Structural Insights into Gene Expression and Protein Synthesis, 2020, pp. 501–511., doi:10.1142/9789811215865_0060.
59. Ban, Nenad, et al. "The Complete Atomic Structure of the Large Ribosomal Subunit at 2.4 Å Resolution." Science, vol. 289, no. 5481, 2000, pp. 905–920., doi:10.1126/science.289.5481.905.
60. Kauffman, Stuart. *At Home in the Universe the Search for the Laws of Self-Organization and Complexity.* Oxford University Press, USA, 2014.
61. Hordijk, Wim. "Complexity Research." *John Templeton Foundation,* www.templeton.org/discoveries/complexity. Accessed online 8/27/2023.

Chapter 10

1. Daròs, José-Antonio, et al. "Viroids: An Ariadne's Thread into the RNA Labyrinth." *EMBO Reports*, vol. 7, no. 6, 2006, pp. 593–598., doi:10.1038/sj.embor.7400706.

2. Flores, Ricardo, et al. "Viroids: Survivors from the RNA World?" *Annual Review of Microbiology*, vol. 68, no. 1, 2014, pp. 395–414., doi:10.1146/annurev-micro-091313-103416.
3. Luigi, M., et al. "Natural spread and molecular analysis of pospiviroids infecting ornamentals in Italy." *Journal of Plant Pathology*. 2011 : 491-495.
4. Perreault, Jean-Pierre, and Martin Pelchat. "Viroids and Hepatitis Delta Virus." *Fundamentals of Molecular Virology*, by Nicholas H. Acheson, John Wiley & Sons, 2011, pp. 378–386.
5. Biondi, Elisa, et al. "Catalytic Activity of Hammerhead Ribozymes in a Clay Mineral Environment: Implications for the RNA World." *Gene*, vol. 389, no. 1, 2007, pp. 10–18., doi:10.1016/j.gene.2006.09.002.
6. Roossinck, M J, et al. "Satellite RNAS of Plant Viruses: Structures and Biological Effects." *Microbiological Reviews*, vol. 56, no. 2, 1992, pp. 265–279., doi:10.1128/mr.56.2.265-279.1992.
7. Prody, Gerry A., et al. "Autolytic Processing of Dimeric Plant Virus Satellite RNA." *Science*, vol. 231, no. 4745, 1986, pp. 1577–1580., doi:10.1126/science.231.4745.1577.
8. Symons, R. H., and J. W. Randles. "Encapsidated Circular Viroid-like Satellite RNAS (Virusoids) of Plants." *Current Topics in Microbiology and Immunology*, 1999, pp. 81–105., doi:10.1007/978-3-662-09796-0_5.
9. Cavalier-Smith, T. "The Neomuran Origin of Archaebacteria, the Negibacterial Root of the Universal Tree and Bacterial Megaclassification." *International Journal of Systematic and Evolutionary Microbiology*, vol. 52, no. 1, 2002, p. 9., doi:10.1099/00207713-52-1-7.
10. Ulbert, Sebastian, et al. "Direct Membrane Protein–DNA Interactions Required Early in Nuclear Envelope Assembly." *Journal of Cell Biology*, vol. 173, no. 4, 2006, pp. 469–476., doi:10.1083/jcb.200512078.
11. Rubino, L., et al. "Nucleotide Sequence and Structural Analysis of Two Satellite RNAS Associated with Chicory Yellow Mottle Virus." *Journal of General Virology*, vol. 71, no. 9, 1990, pp. 1897–1903., doi:10.1099/0022-1317-71-9-1897.

12. Macnaughton, T. B., and M. M. Lai. "HDV RNA Replication: Ancient Relic or Primer?" *Hepatitis Delta Virus,* by J. L. Casey, Springer, 2006, pp. 25–45.
13. Taylor, J. M. "Structure and Replication of Hepatitis Delta Virus RNA." *Hepatitis Delta Virus,* by J. L. Casey, Springer, 2006, p. 2.
14. Zuccola, Harmon J, et al. "Structural Basis of the Oligomerization of Hepatitis Delta Antigen." *Structure,* vol. 6, no. 7, 1998, pp. 821–830., doi:10.1016/s0969-2126(98)00084-7.
15. Lin, H P, et al. "Localization of Isoprenylated Antigen of Hepatitis Delta Virus by Anti-Farnesyl Antibodies." *Journal of General Virology,* vol. 80, no. 1, 1999, pp. 91–96., doi:10.1099/0022-1317-80-1-91.
16. Hemmer, O., et al. "Comparison of the Nucleotide Sequences of Five Tomato Black Ring Virus Satellite RNAS." *Journal of General Virology,* vol. 68, no. 7, 1987, p. 1831., doi:10.1099/0022-1317-68-7-1823.
17. Poisson, F., et al. "Characterization of RNA-Binding Domains of Hepatitis Delta Antigen." *Journal of General Virology,* vol. 74, no. 11, 1993, pp. 2473–2478., doi:10.1099/0022-1317-74-11-2473.
18. Hager, Alicia J., et al. "Ribozymes: Aiming at RNA Replication and Protein Synthesis." *Chemistry & Biology,* vol. 3, no. 9, 1996, pp. 717–725., doi:10.1016/s1074-5521(96)90246-x.
19. Burma, D. P., et al. "Do Ribosomal RNAS Act Merely as Scaffold for Ribosomal Proteins?" *Journal of Biosciences,* vol. 8, no. 3-4, 1985, pp. 757–766., doi:10.1007/bf02702774.
20. G., Theng B K. *Clay Mineral Catalysis of Organic Reactions.* CRC Press, 2019, pp: 399-401
21. Cech, Thomas R. "The Ribosome Is a Ribozyme." *Science,* vol. 289, no. 5481, 2000, pp. 878–879., doi:10.1126/science.289.5481.878.
22. Nissen, Poul, et al. "The Structural Basis of Ribosome Activity in Peptide Bond Synthesis." *Structural Insights into Gene Expression and Protein Synthesis,* 2020, pp. 501–511., doi:10.1142/9789811215865_0060.
23. Ban, Nenad, et al. "The Complete Atomic Structure of the Large Ribosomal Subunit at 2.4 Å Resolution." *Science,* vol. 289, no. 5481, 2000, pp. 905–920., doi:10.1126/science.289.5481.905.

24. Noller, Harry F., et al. "Unusual Resistance of Peptidyl Transferase to Protein Extraction Procedures." *Science*, vol. 256, no. 5062, 1992, pp. 1416–1419., doi:10.1126/science.1604315.
25. Tamura, Koji. "Ribosome Evolution: Emergence of Peptide Synthesis Machinery." *Journal of Biosciences*, vol. 36, no. 5, 2011, pp. 921–928., doi:10.1007/s12038-011-9158-2.
26. Polacek, Norbert, and Alexander S. Mankin. "The Ribosomal Peptidyl Transferase Center: Structure, Function, Evolution, Inhibition." *Critical Reviews in Biochemistry and Molecular Biology*, vol. 40, no. 5, 2005, pp. 285–311., doi:10.1080/10409230500326334.
27. Turon-Lagot, Vincent, et al. "Targeting the Host for New Therapeutic Perspectives in Hepatitis D." *Journal of Clinical Medicine*, vol. 9, no. 1, 2020, p. 222., doi:10.3390/jcm9010222.
28. Alberts, Bruce. "The Lipid Bilayer." *Molecular Biology of the Cell. 4th Edition.*, U.S. National Library of Medicine, 1 Jan. 1970. https://ncbi.nlm.nih.gov/books/NBK26871/. Accessed online on 5/18/2021.
29. Berg, Jeremy M. "There Are Three Common Types of Membrane Lipids." *Biochemistry. 5th Edition.*, U.S. National Library of Medicine, 1 Jan. 1970. https://ncbi.nlm.nih.gov/books/NBK22361/. Accessed online on 5/18/2021.
30. Sohlenkamp, Christian, and Otto Geiger. "Bacterial Membrane Lipids: Diversity in Structures and Pathways." *FEMS Microbiology Reviews*, vol. 40, no. 1, 2015, pp. 133–159., doi:10.1093/femsre/fuv008.
31. Rhoades, Rodney, and David R. Bell. *Medical Physiology: Principles for Clinical Medicine.* Wolters Kluwer Health/Lippincott Williams & Wilkins, 2013, p. 25
32. Lodish, Harvey. *Molecular Cell Biology.* W. H. Freeman, 2016, pp. 271-299.
33. Fan, Jianjun, et al. "Membrane Asymmetry and Phospholipid Translocases in Eukaryotic Cells." *Advances in Membrane Proteins: Mass Processing and Transportation*, SPRINGER, 2018, pp. 47–76.
34. Schaechter, Moselio, et al. *Schaechter's Mechanisms of Microbial Disease.* Wolters Kluwer Health/Lippincott Williams & Wilkins, 2013, p. 21

Chapter 11

1. Avina-Padilla, Katia, et al. "In Silico Prediction and Validation of Potential Gene Targets for Pospiviroid-Derived Small RNAS during Tomato Infection." *Gene*, vol. 564, no. 2, 2015, pp. 197–205., doi:10.1016/j.gene.2015.03.076.
2. Koornneef, Maarten, and David Meinke. "The Development of Arabidopsis as a Model Plant." *The Plant Journal*, vol. 61, no. 6, 2010, pp. 909–921., doi:10.1111/j.1365-313x.2009.04086.x.
3. Diener, T. O. "Are Viroids Escaped Introns?" *Proceedings of the National Academy of Sciences*, vol. 78, no. 8, 1981, pp. 5014–5015., doi:10.1073/pnas.78.8.5014.
4. Lewin, Roger. "Viroids May Be Escaped Introns." *Science*, vol. 233, no. 4771, 1986, pp. 1384–1384., doi:10.1126/science.3749883.
5. Dinter-Gottlieb, G. "Viroids and Virusoids Are Related to Group I Introns." *Proceedings of the National Academy of Sciences*, vol. 83, no. 17, 1986, pp. 6250–6254., doi:10.1073/pnas.83.17.6250.
6. Nielsen, Henrik, and Steinar D. Johansen. "Group I Introns: Moving in New Directions." *RNA Biology*, vol. 6, no. 4, 2009, pp. 375–383., doi:10.4161/rna.6.4.9334.
7. Francis, Warren R., and Gert Wörheide. "Similar Ratios of Introns to Intergenic Sequence across Animal Genomes." *Genome Biology and Evolution*, vol. 9, no. 6, 2017, pp. 1582–1598., doi:10.1093/gbe/evx103.
8. Schmidt-Puchta, Waltraud, et al. "Nucleotide Sequence of the Intergenic Spacer (IGS) of the Tomato Ribosomal DNA." *Plant Molecular Biology*, vol. 13, no. 2, 1989, pp. 251–253., doi:10.1007/bf00016143.
9. Siguier, Patricia, et al. "Insertion Sequences in Prokaryotic Genomes." *Current Opinion in Microbiology*, vol. 9, no. 5, 2006, pp. 526–531., doi:10.1016/j.mib.2006.08.005.
10. Delihas, Nicholas. "Intergenic Regions of Borrelia Plasmids Contain Phylogenetically Conserved RNA Secondary Structure Motifs." *BMC Genomics*, vol. 10, no. 1, 2009, p. 101., doi:10.1186/1471-2164-10-101.

11. Touchon, Marie, and Eduardo P. Rocha. "Causes of Insertion Sequences Abundance in Prokaryotic Genomes." *Molecular Biology and Evolution*, vol. 24, no. 4, 2007, pp. 969–981., doi:10.1093/molbev/msm014.
12. Silva, J. C., et al. "Conserved Fragments of Transposable Elements in Intergenic Regions: Evidence for Widespread Recruitment of Mir- and L2-Derived Sequences within the Mouse and Human Genomes." *Genetical Research*, vol. 82, no. 1, 2003, pp. 1–18., doi:10.1017/s0016672303006268.
13. Kiefer, M. C., et al. "Structural Similarities between Viroids and Transposable Genetic Elements." *Proceedings of the National Academy of Sciences*, vol. 80, no. 20, 1983, pp. 6234–6238., doi:10.1073/pnas.80.20.6234.
14. Varmus, Harold E. "Form and Function of Retroviral Proviruses." *Science*, vol. 216, no. 4548, 1982, pp. 812–820., doi:10.1126/science.6177038.

Chapter 12

1. Van Duin, Jan. "Single-Stranded RNA Bacteriophages ." *Fundamentals of Molecular Virology*, by Nicholas H. Acheson, John Wiley & Sons, 2011, p. 59.
2. Shiba, Tadayoshi, and Yurie Suzuki. "Localization of a Protein in the RNA-a Protein Complex of RNA Phage MS2." *Biochimica Et Biophysica Acta (BBA) - Nucleic Acids and Protein Synthesis*, vol. 654, no. 2, 1981, pp. 249–255., doi:10.1016/0005-2787(81)90179-9.
3. Mindich, Leonard. "Phages with Segmented Double-Stranded RNA Genomes." *The Bacteriophages*, by Richard Calendar, Oxford University Press, 2006, p. 197.
4. Olkkonen, V. M., et al. "In Vitro Assembly of Infectious Nucleocapsids of Bacteriophage Phi 6: Formation of a Recombinant Double-Stranded RNA Virus." *Proceedings of the National Academy of Sciences*, vol. 87, no. 23, 1990, pp. 9173–9177., doi:10.1073/pnas.87.23.9173.
5. Mindich, Leonard. "Precise Packaging of the Three Genomic Segments of the Double-Stranded-RNA Bacteriophage ϕ6." *Microbiology*

and Molecular Biology Reviews, vol. 63, no. 1, 1999, pp. 149–160., doi:10.1128/mmbr.63.1.149-160.1999.

6. "Complementary DNA." *Wikipedia*, Wikimedia Foundation, https://en.wikipedia.org/wiki/Complementary_DNA. Accessed online on 11/20/2020.
7. "Complementary DNA." *Biology Articles, Tutorials & Dictionary Online*, https://biologyonline.com/dictionary/complementary-dna. Accessed online on 11/20/2020.
8. Nemecek, D., et al. "Packaging Accessory Protein P7 and Polymerase P2 Have Mutually Occluding Binding Sites inside the Bacteriophage 6 Procapsid." *Journal of Virology*, vol. 86, no. 21, 2012, pp. 11616–11624., doi:10.1128/jvi.01347-12.
9. Ojala, P M, et al. "Protein P4 of Double-Stranded RNA Bacteriophage Phi 6 Is Accessible on the Nucleocapsid Surface: Epitope Mapping and Orientation of the Protein." *Journal of Virology*, vol. 67, no. 5, 1993, pp. 2879–2886., doi:10.1128/jvi.67.5.2879-2886.1993.

Chapter 13

1. Ackermann, Hans-W. "Classification of Bacteriophages." *The Bacteriophages*, by Richard Calendar, Oxford University Press, 2006, p. 10.
2. Ackerman, Hans-W. "Bacteriophage Classification." *Bacteriophages: Biology and Applications*, by Elizabeth Kutter and Alexander Sulakvelidze, CRC Press, 2005, p. 72.
3. Lodish, Harvey F. *Molecular Cell Biology*. W.H. Freeman-Macmillan Learning, 2016, p. 423
4. Hesselberth, Jay R. "Lives That Introns Lead after Splicing." *Wiley Interdisciplinary Reviews: RNA*, vol. 4, no. 6, 2013, pp. 677–691., doi:10.1002/wrna.1187.
5. Clement, Jade Q., et al. "The Stability and Fate of a Spliced Intron from Vertebrate Cells." *RNA*, vol. 5, no. 2, 1999, pp. 206–220., doi:10.1017/s1355838299981190.
6. Harrison, Stephen C. "Virus Structure and Assembly." *Fundamentals of Molecular Virology*, by Nicholas H. Acheson, John Wiley & Sons, 2011, p. 28.

7. Murray, Patrick R., et al. *Medical Microbiology.* Elsevier, 2016, p. 533.
8. Krupovic, Mart, et al. "Viruses of Archaea: Structural, Functional, Environmental and Evolutionary Genomics." *Virus Research,* vol. 244, 2018, pp. 181–193., doi:10.1016/j.virusres.2017.11.025.
9. Colson, Philippe, et al. "'Megavirales', a Proposed New Order for Eukaryotic Nucleocytoplasmic Large DNA Viruses." *Archives of Virology,* vol. 158, no. 12, 2013, pp. 2517–2521., doi:10.1007/s00705-013-1768-6.
10. Iyer, Lakshminarayan M., et al. "Common Origin of Four Diverse Families of Large Eukaryotic DNA Viruses." *Journal of Virology,* vol. 75, no. 23, 2001, pp. 11720–11734., doi:10.1128/jvi.75.23.11720-11734.2001.
11. Aherfi, Sarah, et al. "Giant Viruses of Amoebas: An Update." *Frontiers in Microbiology,* vol. 7, 2016, doi:10.3389/fmicb.2016.00349.
12. Mutsafi, Yael, et al. "Membrane Assembly during the Infection Cycle of the Giant Mimivirus." *PLoS Pathogens,* vol. 9, no. 5, 2013, doi:10.1371/journal.ppat.1003367.
13. Scola, Bernard La, et al. "A Giant Virus in Amoebae." *Science,* vol. 299, no. 5615, 2003, pp. 2033–2033., doi:10.1126/science.1081867.
14. Raoult, Didier, et al. "The 1.2-Megabase Genome Sequence of Mimivirus." *Science,* vol. 306, no. 5700, 2004, pp. 1344–1350., doi:10.1126/science.1101485.
15. Raoult, Didier, and Patrick Forterre. "Redefining Viruses: Lessons from Mimivirus." *Nature Reviews Microbiology,* vol. 6, no. 4, 2008, pp. 315–319., doi:10.1038/nrmicro1858.
16. Cordingley, Michael G. *Viruses: Agents of Evolutionary Invention.* Harvard University Press, 2017, pp. 159-166.
17. Fermin, Gustavo, et al. "Viruses of Prokaryotes, Protozoa, Fungi, and Chromista." *Viruses: Molecular Biology, Host Interactions and Applications to Biotechnology,* by Paula Tennant et al., Academic Press, 2018, p. 234.
18. Moelling, Karin. *Viruses: More Friends than Foes.* World Scientific, 2017, pp. 140-141

19. Boyer, M., et al. "Giant Marseillevirus Highlights the Role of Amoebae as a Melting Pot in Emergence of Chimeric Microorganisms." *Proceedings of the National Academy of Sciences*, vol. 106, no. 51, 2009, pp. 21848–21853., doi:10.1073/pnas.0911354106.
20. Koonin, Eugene V., et al. "Evolution of Double-Stranded DNA Viruses of Eukaryotes: From Bacteriophages to Transposons to Giant Viruses." *Annals of the New York Academy of Sciences*, vol. 1341, no. 1, 2015, pp. 10–24., doi:10.1111/nyas.12728.
21. "Virus Structure." *Michigan State University*, https://msu.edu/course/mmg/569/Virus%20Structure.htm. Accessed online on 11/20/2020.
22. "Category:Mimivirus." *Wikimedia Commons*. https://commons.wikimedia.org/wiki/Category:Mimivirus. Retrieved on 12/03/2021.

Chapter 14

1. Scola, Bernard La, et al. "A Giant Virus in Amoebae." *Science*, vol. 299, no. 5615, 2003, pp. 2033–2033., doi:10.1126/science.1081867.
2. Xiao, Chuan, et al. "Structural Studies of the Giant Mimivirus." *PLoS Biology*, vol. 7, no. 4, 2009, doi:10.1371/journal.pbio.1000092.
3. Jain, Samta. "Biosynthesis of Archaeal Membrane Ether Lipids." *Frontiers in Microbiology*, vol. 5, 2014, doi:10.3389/fmicb.2014.00641.
4. Paltauf, Fritz. "Ether Lipids in Biomembranes." *Chemistry and Physics of Lipids*, vol. 74, no. 2, 1994, pp. 101–139., doi:10.1016/0009-3084(94)90054-x.
5. Learning, Lumen. "Biology for Majors II." https://courses.lumenlearning.com/wm-biology2/chapter/archaea-vs-bacteria/. Accessed online on 01/27/2021.
6. Koga, Y., et al. "Did Archaeal and Bacterial Cells Arise Independently from Noncellular Precursors? A Hypothesis Stating That the Advent of Membrane Phospholipid with Enantiomeric Glycerophosphate Backbones Caused the Separation of the Two Lines of Descent." *Journal of Molecular Evolution*, vol. 46, no. 1, 1998, pp. 54–63., doi:10.1007/pl00006283.
7. Sohlenkamp, Christian, and Otto Geiger. "Bacterial Membrane Lipids: Diversity in Structures and Pathways." *FEMS Microbiology Reviews*, vol. 40, no. 1, 2015, pp. 133–159., doi:10.1093/femsre/fuv008.

Chapter 15

1. Muller, Richard A. *Now: The Physics of Time: Richard A. Muller*. W.W. Norton & Company, 2016, p. 131.
2. Fleisher, Paul. *The Big Bang*. Lerner, 2006, p. 16.
3. Lemaître, G. "UN Univers Homogène De Masse Constante Et De Rayon Croissant Rendant Compte De La Vitesse Radiale Des Nébuleuses Extra-Galactiques." *NASA/ADS*, 1 Jan. 1970, ui.adsabs.harvard.edu/abs/1927ASSB...47...49L/abstract.
4. Luminet, Jean-Pierre. "Lemaitre's Big Bang." https://arxiv.org/ftp/arxiv/papers/1503/1503.08304.pdf
5. Muller, Richard A. *Now: The Physics of Time: Richard A. Muller*. W.W. Norton & Company, 2016, p. 157.
6. Carroll, Sean M. *From Eternity to Here: The Quest for the Ultimate Theory of Time*. Dutton, 2016, p. 58
7. Lederman, Leon M., and Dick Teresi. *The God Particle*. Bantam, 1993, p. 106.
8. Krauss, Lawrence Maxwell. *A Universe from Nothing: Why There Is Something Rather than Nothing*. Atria Paperback, 2013.
9. Kim, Eunsoo. *Time, Eternity, and the Trinity: A Trinitarian Analogical Understanding of Time and Eternity*. Pickwick Publications, 2010, p. 76.
10. Plato, et al. *Timaeus and Critias*. Oxford University Press, 2008, p. 18.
11. "Nothing." *Wikipedia*, Wikimedia Foundation. https://en.wikipedia.org/wiki/Nothing. Accessed online on 07/22/2021.
12. Hawkings, Steven. *A Brief History of Time*. Bantam Books, 1988, p. 121.
13. Rovelli, Carlo, et al. *Seven Brief Lessons on Physics*. Penguin Books, 2016, p. 46.
14. "Fundamental Physical Constants." *Physical Measurement Laboratory*. https://physics.nist.gov/cgi-bin/cuu/Value?plktmp|search_for=planck+temperature. Accessed online on 03/20/2022.
15. "Brief History of the Universe." *Brief History of the Universe*, https://www.astro.ucla.edu/~wright/BBhistory.html. Accessed online on 01/04/2021.

16. "Wien's Law." *Wien's Law.* http://hosting.astro.cornell.edu/academics/courses/astro201/wiens_law.htm. Accessed online on 01/04/2021.
17. Fleisher, Paul. *The Big Bang.* Lerner, 2006, p. 47
18. "Fundamental Physical Constants." *Physical Measurement Laboratory.* https://physics.nist.gov/cgi-bin/cuu/Value?plkl|search_for=planck+length. Accessed online on 03/20/2022.
19. "Fundamental Physical Constants." *Physical Measurement Laboratory.* https://physics.nist.gov/cgi-bin/cuu/Value?plkt|search_for=planck+time. Accessed online on 03/20/2022.
20. "Planck Units." *Wikipedia*, Wikimedia Foundation. https://en.wikipedia.org/wiki/Planck_units. Retrieved on 08/10/2021.
21. "First Law of Thermodynamics." *Wikipedia*, Wikimedia Foundation. https://en.wikipedia.org/wiki/First_law_of_thermodynamics. Retrieved on 08/10/2021.
22. "1911 Encyclopædia Britannica/Empedocles." *Wikisource, the Free Online Library*, Wikimedia Foundation, Inc., 9 Feb. 2019. https://en.wikisource.org/wiki/1911_Encyclop%C3%A6dia_Britannica/Empedocles. Retrieved on 8/17/2021.
23. Challoner, Jack. *The Atom: A Visual Tour.* The MIT Press, 2018, p. 20.
24. "Aristotle." *Wikipedia*, Wikimedia Foundation. https://en.wikipedia.org/wiki/Aristotle. Retrieved online on 8/17/2021.
25. Challoner, Jack. *The Atom: A Visual Tour.* The MIT Press, 2018, p. 33-34.
26. Lederman, Leon M., and Dick Teresi. *The God Particle: If the Universe Is the Answer, What Is the Question?* Houghton Mifflin, 2006, p. 184.
27. Lederman, Leon M., and Dick Teresi. *The God Particle: If the Universe Is the Answer, What Is the Question?* Houghton Mifflin, 2006, p. 155.
28. Carroll, Sean M. *The Particle at the End of the Universe: How the Hunt for the Higgs Boson Leads Us to the Edge of a New World.* Plume, 2016, pp: 34-35.
29. "The Nature of Radiation." *6(f). The Nature of Radiation.* physicalgeography.net/fundamentals/6f.html. Retrieved on 02/14/2021.

30. "Black-Body Radiation." *Wikipedia*, Wikimedia Foundation. https://en.wikipedia.org/wiki/Black-body_radiation. Retrieved on 2/14/2021.

Chapter 16

1. PONNAMPERUMA, CYRIL, et al. "Synthesis of Adenosine Triphosphate under Possible Primitive Earth Conditions." *Nature*, vol. 199, no. 4890, 1963, pp. 222–226., doi:10.1038/199222a0.
2. Schreiber, Alexander, and Steven Gimbel. "Evolution and the Second Law of Thermodynamics: Effectively Communicating to Non-Technicians." *Evolution: Education and Outreach*, vol. 3, no. 1, 2010, pp. 99–106., doi:10.1007/s12052-009-0195-3.
3. Finnerty, John R., et al. "Origins of Bilateral Symmetry: Hox and DPP Expression in a Sea Anemone." *Science*, vol. 304, no. 5675, 2004, pp. 1335–1337., doi:10.1126/science.1091946.
4. Heinrich, Bernd, and George A. Bartholomew. "Temperature Control in Flying Moths." *Scientific American*, vol. 226, no. 6, 1972, pp. 70–77., doi:10.1038/scientificamerican0672-70.
5. Bakker, Robert T. "Dinosaur Renaissance." *Scientific American*, vol. 232, no. 4, 1975, pp. 58–79., doi:10.1038/scientificamerican0475-58.
6. Harlow, Peter, and Gordon Grigg. "Shivering Thermogenesis in a Brooding Diamond Python, Python Spilotes Spilotes." *Copeia*, vol. 1984, no. 4, 1984, p. 959., doi:10.2307/1445340.
7. Grady, John M., et al. "Evidence for Mesothermy in Dinosaurs." *Science*, vol. 344, no. 6189, 2014, pp. 1268–1272., doi:10.1126/science.1253143.
8. Bennett, Albert F., and John A. Ruben. "Endothermy and Activity in Vertebrates." *Science*, vol. 206, no. 4419, 1979, pp. 649–654., doi:10.1126/science.493968.
9. Voet, Donald, et al. *Fundamentals of Biochemistry: Life at the Molecular Level.* John Wiley & Sons, 2016, p. 624.
10. Karasov, William H., and Jared M. Diamond. "Digestive Adaptations for Fueling the Cost of Endothermy." *Science*, vol. 228, no. 4696, 1985, pp. 202–204., doi:10.1126/science.3975638.

11. Prothero, Donald R., and Robert H. Dott. *Evolution of the Earth*. McGraw-Hill, 2010, p. 398.
12. Rezende, Enrico L., et al. "Shrinking Dinosaurs and the Evolution of Endothermy in Birds." *Science Advances*, vol. 6, no. 1, 2020, doi:10.1126/sciadv.aaw4486.
13. Berna, F., et al. "Microstratigraphic Evidence of in Situ Fire in the Acheulean Strata of Wonderwerk Cave, Northern Cape Province, South Africa." *Proceedings of the National Academy of Sciences*, vol. 109, no. 20, 2012, doi:10.1073/pnas.1117620109.
14. James, Steven R., et al. "Hominid Use of Fire in the Lower and Middle Pleistocene: A Review of the Evidence [and Comments and Replies]." *Current Anthropology*, vol. 30, no. 1, 1989, pp. 1–26., doi:10.1086/203705.
15. Papagianni, Dimitra, and Michael Ari Morse. *The Neanderthals Rediscovered How Modern Science Is Rewriting Their History*. Thames & Hudson, 2015, p. 83
16. Darwin, Charles. *The Descent of Man*. Revised Edition ed., Merrill and Baker, 1900, p. 54.
17. "Atlantic Slave Trade." *Wikipedia*, Wikimedia Foundation, en.wikipedia.org/wiki/Atlantic_slave_trade. Retrieved on 08/22/2021.
18. Magazine, Smithsonian. "Why Fire Makes Us Human." *Smithsonian.com*, Smithsonian Institution, 1 June 2013. https://www.smithsonianmag.com/science-nature/why-fire-makes-us-human-72989884/. Accessed online on 11/27/2021.
19. Zebrowski, Ernest. *Global Climate Change*. Imagine, 2011, p. 156.
20. Zebrowski, Ernest. *Global Climate Change*. Imagine, 2011, p. 98.

Epilogue

1. Rupke, Nicolaas A. *Richard Owen Biology without Darwin*. University of Chicago Press, 2009, p. 173.
2. Gaarder Haug, Espen. "The Gravitational Constant and the Planck Units. A Simplification of the Quantum Realm." *Physics Essays*, vol. 29, no. 4, 2016, pp. 558–561., doi:10.4006/0836-1398-29.4.558.
3. Gould, Stephen Jay. *Wonderful Life: The Burgess Shale and the Nature of History*. Norton, 2007, pp: 321-323.

4. Vignuzzi, Marco, et al. "Quasispecies Diversity Determines Pathogenesis through Cooperative Interactions in a Viral Population." *Nature*, vol. 439, no. 7074, 2005, pp. 344–348., doi:10.1038/nature04388.
5. Lauring, Adam S., and Raul Andino. "Quasispecies Theory and the Behavior of RNA Viruses." *PLoS Pathogens*, vol. 6, no. 7, 2010, doi:10.1371/journal.ppat.1001005.
6. President John F. Kennedy. "Excerpt from the 'Special Message to the Congress on Urgent National Needs'." May 25, 1961. https://www.nasa.gov/vision/space/features/jfk_speech_text.html
7. Hofstadter, Richard, and Eric Foner. *Social Darwinism in American Thought*. Beacon Press, 2021, p. 97.
8. Margulis, Lynn, and Dorion Sagan. *Microcosmos: Four Billion Years of Evolution from Our Microbial Ancestors*. University of California Press, 1998.
9. Martins, Zita. "The Nitrogen Heterocycle Content of Meteorites and Their Significance for the Origin of Life." *Life*, vol. 8, no. 3, 2018, p. 28., doi:10.3390/life8030028.
10. Glavin, D P, and J L Bada. *Isolation of Purines and Pyrimidines from the Murchison Meteorite Using Sublimation*. https://www.lpi.usra.edu/meetings/lpsc2004/pdf/1022.pdf. Accessed online on 03/20/2022.
11. Rotelli, Luca, et al. "The Key Role of Meteorites in the Formation of Relevant Prebiotic Molecules in a Formamide/Water Environment." *Scientific Reports*, vol. 6, no. 1, 2016, doi:10.1038/srep38888.
12. Saladino, Raffaele, et al. "Meteorite-Catalyzed Syntheses of Nucleosides and of Other Prebiotic Compounds from Formamide under Proton Irradiation." *Proceedings of the National Academy of Sciences*, vol. 112, no. 21, 2015, doi:10.1073/pnas.1422225112.
13. Darwin, Charles. *The Descent of Man*. Revised Edition ed., Merrill and Baker, 1900, p. 707.
14. Kneale, M. "XIII—Eternity and Sempiternity." *Proceedings of the Aristotelian Society*, vol. 69, no. 1, 1969, pp. 223–238., doi:10.1093/aristotelian/69.1.223.
15. Plato. "Timaeus." *The Project Gutenberg EBook of Timaeus*. https://www.gutenberg.org/files/1572/1572-h/1572-h.htm
16. Augustine of Hippo, Confessions, tr. R. S. Pine-Coffin. (London 1961), p. 263.

Index

Note: Page numbers in *italics* indicate a figure and page numbers in **bold** indicate a table on the corresponding page. Page numbers followed by "n" refer to notes.

C

H

I

N

Q

S

Made in the USA
Middletown, DE
13 November 2023

42561074R00198